T0133388

Nuclear Rivals

Anglo-American Atomic Relations, 1941–1952

Septimus H. Paul

Ohio State University Press

Columbus

Library of Congress Cataloging-in-Publication Data

Paul, Septimus H.
    Nuclear rivals : Anglo-American atomic relations, 1941–1952 / Septimus
H. Paul.
        p. cm.
Includes bibliographical references and index.
    ISBN 0-8142-0852-5 (cloth : alk. paper)
    1. Nuclear weapons—United States.   2. Nuclear weapons—Great
Britain.   3. United States—Military relations—Great Britain.   4. Great
Britain—Military relations—United States.   I. Title.
    UA23 .P3724 2000
    355.02′17′097309044—dc21

00-008677

Type set in Times Roman by Graphic Composition, Inc.
Printed by Thomson-Shore, Inc.

The paper used in this publication meets the minimum requirements of the
American National Standard for Information Sciences—Permanence of
Paper for Printed Library Materials. ANSI Z39.48–1992.

9 8 7 6 5 4 3 2 1

This work is dedicated to Jo-Ann, my wife, who helped me make the transition from the typewriter to the computer.

Contents

Acknowledgments

I am grateful to the following scholars for the help they provided: Professors Bentley B. Gilbert, Robert L. Messer, Marion S. Miller, James Sack, and Eugene Beiriger. Their guidance helped me weed out irrelevancies in the manuscript and sharpen the focus of my analyses. In particular, I am indebted to Professor Gilbert, whose diligent reading of the early drafts provided constructive criticisms and helpful suggestions. His expertise in British history was an invaluable asset. Professor Messer also helped me immeasurably. It was he who drew my attention to the myriad of unanswered questions surrounding Anglo-American atomic relations; I also benefited from his knowledge of American atomic diplomacy. I thank also the readers selected by the Ohio State University Press. Their insightful comments strengthened the final draft.

I appreciate the help provided by the staff members of the following depositories: the National Archives in Washington D.C.; the Truman Library in Missouri; the Library of Congress; the University of Illinois (Chicago) Library; the Public Records Office in London; Churchill College, Cambridge; the Bodleian Library, Oxford; and the British Library of Economics. A special thanks to Mr. Dennis Bilger, the archivist at the Truman Library, who identified some relevant sources on deposit at the library.

Introduction

In 1945, the United States developed the atomic bomb with the collaboration of the United Kingdom and a host of scientists from around the world. The British later claimed that the United States had led them to believe that this collaboration, under the terms of the wartime Quebec Agreement (1943) and the Hyde Park Aide-Memoire (1944), would continue after the war until, at the earliest, the British had developed their own bomb. At the conclusion of the war, the United States did not honor fully this promise or agreement. The American administration did not deny the existence of some sort of wartime agreement but did question its postwar viability. Nevertheless, the administration continued to lead the British to believe that the terms of the agreement, if not the agreement itself, would be honored. Negotiations dragged on for seven years, during which time the Americans made vague written and spoken promises that the British accepted, only to have them broken by the Americans. In the interim, the British began their own atomic program, which ultimately succeeded in developing an atomic bomb in October 1952. This book will examine the course of Anglo-American atomic energy cooperation during this critical period in the history of the British atomic program.

From the inception of the Anglo-American atomic relationship, during the war years, mistrust and controversies plagued the partnership. Members of the U.S. atomic energy team who opposed all efforts to work with the British created these problems. Ironically, before the partnership was formalized, the Americans had initiated the idea of collaboration as early as 1941. At that stage, the British program was far more advanced than that of the United States. Hence, the British were cold to the idea. By the end of 1942, several factors surfaced to force the British to reappraise their opposition to a joint effort. It was clear that American atomic research had crept ahead of them. Moreover, they were forced to consider the security, financial, and technical challenges of setting up a project in Britain. Meanwhile, the Americans took a second look at the combined effort they had once favored and decided that an independent American program best served the national interest. President

Franklin Roosevelt of the United States agreed. Nevertheless, thanks to the dauntless efforts of Prime Minister Winston Churchill of the United Kingdom and the exigencies of World War II, Roosevelt was forced to change his policy in August 1943 by agreeing to atomic collaboration with the British.

Despite Roosevelt's acquiescence, his closest atomic advisers continued to erect roadblocks in order to maintain American dominance over atomic research. Vannevar Bush and James Conant, his earliest advisers, were quick to grasp the enormous military, political, and economic potential of atomic power in the postwar world and thus resisted collaboration. They were averse to sharing atomic information with any other nation. Despite their strongly held views, they had no choice but to adopt the president's policy. This study makes clear that the policy of noncollaboration was introduced and put in place by Bush and Conant, not by General Leslie Groves, director of the Manhattan Project. Groves efficiently carried out a policy that he had inherited and for which he believed he had the support of the president. When Roosevelt appeared to undermine this support by forcing the Quebec Agreement upon him, Groves was none too happy. He accepted the British scientists into the Manhattan Project but under his terms. Under the guise of the policy of compartmentalization, he limited their access to vital bomb-making information. As a result, after the war ended, these scientists returned to the United Kingdom with major gaps in their overall knowledge of bomb production. Meanwhile, however, Churchill and Roosevelt, in September 1944, had already sown the seeds of future controversy by secretly agreeing to continue postwar atomic cooperation. The decision was made with no input from the president's advisers. This so-called Hyde Park Aide-Memoire came to light in the United States only after Roosevelt's death.

At the conclusion of the war, the British expected atomic collaboration to continue. Specifically, they expected the Americans to help them develop an atomic bomb, or, at the least, entrust them with their quota of a few bombs. These expectations were never realized. The history of what little collaboration there was between 1945 through 1952 was dictated invariably by the national security interest of the United States, not by any effort on the part of the United States to help develop a U.K. atomic program. An expansion of the U.S. atomic program and a projected shortage of uranium preceded every instance of American concession. As both countries jointly controlled the known sources of uranium, the United States, in times of expansion or shortages, was forced to appease the British through a limited offer of scientific, as opposed to technical, information in order to gain a larger allocation of uranium. Forces outside of the administration were firm in their determination to thwart the British in their efforts to construct a competing atomic program in the United Kingdom, a program that would have consumed a part of the limited supply of available uranium. The most effective way of achieving this aim was to deprive the British of the technical knowledge required to construct an atomic bomb.

President Harry Truman, Secretary of State Dean Acheson, and David Lilienthal, chairman of the Atomic Energy Commission (AEC), reluctantly adopted the policy of noncollaboration because they were unable to work around congressional and military opposition to find a way to share the "secret" with the British. From the outset, President Truman was quite willing to collaborate with the British and would have done so were it not for the efforts of General Groves in 1946. Groves's role turned out to be pivotal in terms of redefining Truman's early postwar inclination toward atomic collaboration. Groves outlined to the administration the pitfalls of sharing atomic data with the British; he leaked security information to the media, and he established personal contacts with Congress in order to apply legislative pressures on the executive branch—all in the name of protecting the "secret" of the bomb. In 1948, the administration and the AEC made another attempt to share atomic information with the British. Again they were thwarted, this time by the efforts of Senators Hickenlooper and Vandenburg, two members of the Joint Congressional Committee on Atomic Energy (JCAE), and Commissioner Lewis Strauss of the AEC. The following year, Truman was prepared fully to bypass existing American laws to institute full atomic cooperation but again was forced to retreat in the face of congressional opposition.

Both Acheson and Lilienthal supported atomic collaboration. Lilienthal saw the production of atomic bombs as one of his primary responsibilities. He recognized also that the production of bombs required a reliable source of raw materials. Uranium could not be guaranteed without closer atomic ties with the British. For his part, Acheson accepted the desirability of closer atomic ties, not only because the British were the Americans' closest allies but also because he believed sincerely that the wartime commitments should be honored. In addition, he could not reconcile Anglo-American cooperation around the world with the absence of such collaboration in the isolated field of atomic energy.

One of the ironies of this drawn-out controversy was that it took place against the backdrop of a close Anglo-American cooperative effort to contain worldwide Soviet expansion. The period from 1945 to 1952 saw the creation of Soviet-style communist regimes throughout Eastern Europe and the proliferation of communist movements in the Third World. The alliance between the United States and the United Kingdom tightened in the immediate postwar years as the two nations drew closer to combat the perceived threat of communism. As the Cold War intensified, the governments of Prime Minister Clement Attlee and President Truman found common ground as they opposed communist activities in Eastern Europe, China, Korea, and Indochina. However, in the area of atomic collaboration this partnership was nonexistent. Voices of opposition arose from individuals in Congress, the military, and even the AEC, offering myriad reasons why the awesome power of atomic energy should not be shared with other powers, regardless of their close association with the United States. Like Bush and Conant before them, the new opponents of collaboration recognized that the monopoly of atomic weapons provided the United States with unsurpassed superpower status. Its secret must be guarded, even at the risk of creating tensions with U.S. allies.

Try as it might, the administration could not circumvent the articles of the McMahon Act that legislated against sharing atomic information with foreign countries. This entire episode of Anglo-American atomic cooperation is a fascinating example of an administration being forced by elements it could not control to adopt a foreign policy it did not support fully. This foreign policy of noncollaboration with the British was imposed upon it by the JCAE, which believed it was exercising its mandate, provided by the McMahon Act, to protect the atomic "secret." The president might have the constitutional powers to conduct foreign policy but not if that policy included sharing atomic information with foreign powers. This question of executive versus congressional power might have had constitutional repercussions if the president had not retreated.

This study also serves as a lucid illustration of the role of modern bureaucracies in the formulation of government policies. From the inception of the Anglo-American atomic energy debate, the intrusion of bureaucrats, more than the influence of government leaders, charted the course of atomic policies. This bureaucratic influence can be identified particularly in the United States. During the war years, Bush, Conant, and Groves, despite the best intent of their president and the opposition of the American atomic scientists, imposed the policy of partial collaboration upon the British. This theme prevails again in the postwar period, when U.S. atomic policies evolved out of recurrent departmental and interdepartmental discussions. The president's role was rather peripheral.

Among the new interpretations and findings, some of which have already been mentioned, are those concerning the effect of the spy scandals of 1950 and 1951. The evidence contradicts the accepted view that the Klaus Fuchs spy scandal of early 1950 destroyed the chances of atomic collaboration. On the contrary, the United States still was eager to cooperate after the scandal in order to win the United Kingdom's assistance in cornering the world's supply of uranium to feed its expanded bomb program. It was the Maclean-Burgess scandal, eighteen months later, that adversely affected the movement toward atomic integration. It was only then that the Americans concluded that the security breaches in the British atomic project could undermine the security of the American program.

This study also made the surprising discovery that the United States, more than the United Kingdom, was responsible for keeping alive the idea of atomic cooperation in the postwar period. It turned out that the United States had as much to gain from collaboration as the British, or more. For example, it is quite likely that all talk of atomic collaboration would have ended in 1946 if the Americans had not taken steps the following year to encourage the British to resurrect it. The British reaction to American resistance to the exchange of bomb information, in 1946, had been swift and decisive. They demanded their fair share of the Congo ores, which they and the Americans jointly controlled. The Americans grudgingly agreed, and this ore, 50 percent of the Congo output, was stockpiled in Britain. Then, in January 1947, the British government, with the utmost secrecy and despite its financial and economic disability, made the decision to establish an independent bomb project. In that same year, the U.S. AEC projected a uranium shortfall for 1948 and beyond—a shortfall that could be prevented only if the British were persuaded to make their own stockpile of uranium and the entire Congo output available to the United States. To win over the British, the United States offered to reopen negotiations on atomic energy cooperation. This negotiation culminated, in January 1948, in the signing of the Modus Vivendi, a loosely worded document that provided the United States with the uranium it coveted and the British with more frustrations over unfulfilled atomic exchanges. The Modus collapsed within months.

The year 1949 opened with another U.S. initiative to revive collaboration. Secretary of State Dean Acheson, for one, saw the need to broaden the scope of Anglo-American foreign relations to include atomic energy matters. However, the overriding incentive for this new approach was the AEC's concern over the never-ending problem of uranium shortages. The administration won over the military and, after a near constitutional crisis, the JCAE and other congressional leaders. The British were passive bystanders during most of this venture. In the end, however, the spy scandals derailed the prospect of an agreement.

After the spy scandals the United States lost interest in atomic collaboration. The British had become dispensable in the uranium equation. Technological factors, research, and exploration had placed the United States in a position to solve its raw material problem. Moreover, it was abundantly clear to the Americans that the United Kingdom, in addition to having security breaches, had declined as a world power. Under the Labour government the British suffered one major foreign policy setback after another. By 1950, the United States had clearly lost what little respect it still had for the British government. The Americans saw no need to share their atomic secrets with a spent power, especially since they were well positioned to fill the vacuum created by British decline. The Americans bluntly conveyed to the British the consensus prevailing in U.S. circles that the British had to earn the right to equal atomic status by developing their own atomic facilities.

In the final analysis, the Americans gained more from the loose partnership that existed up to 1952. Their bomb program used the entire output of the Congo uranium mines, except for a brief period in 1946 and 1947. On the other hand, the British had little to show for their concessions. The Labour government, to its credit, did not sit idly by waiting for the infusion of U.S. atomic information. In January 1947, it began the long process of starting up a bomb program. In the face of financial problems, the daunting task of building a socialist state, and the challenges of ruling a colonial empire, the Labour government found the funds for its atomic program. It could not afford not to have one, especially since it knew that the Soviet Union too was in pursuit of atomic power. Regardless of the cost, it had to possess, and be able to construct, atomic bombs, that new symbol of military power. Thus, as negotiations with the Americans dragged on, British scientists duplicated atomic work already done by American scientists in order to resolve technical problems that the Americans had long resolved.

The year 1949 was an eventful one for the British program. The Americans presented them with a proposal to integrate the U.S. and U.K. atomic programs in exchange for a quota of U.S. atomic bombs. The British response was mild, since the proposal also called for them to get out of the bomb-making business. Before serious negotiations could begin, they and the Americans made the startling discovery that the Soviets had joined the atomic club. This development forced the British government to reevaluate its independent deterrence policy. No amount of British fortitude, unsupported by a stockpile of atomic weapons, would stop Soviet atomic bombs from crossing the English Channel. For a nation that had ruled from the top for so many centuries, it was difficult to consider placing its security in the hands of the United States. But that is exactly what it was prepared to do before the spy scandals broke, derailing the integration movement and slowing the atomic collaboration momentum.

The British effort to construct a bomb would not go to waste. It came to fruition in 1952. In the end, the British had to rely on their own resources to produce an atomic bomb. This study ends with their development of an atomic bomb, but the collaboration problem did not end in 1952. The achievement of the British failed to win them acceptance by the Americans, who, a few weeks after the British detonation, entered the hydrogen age. The "watchdogs" had still another secret to protect. Two years later, in 1954, the Soviets too tested a thermonuclear device. Full atomic collaboration would not come until 1958, after the British tested its own hydrogen bomb.

ONE

The United Kingdom Rejects Early Wartime Collaboration: 1939–1942

In December 1938, Otto Hahn and Fritz Strassman, two German chemists, discovered nuclear fission. Almost immediately after the discovery of fission, the governments of both the United States and the United Kingdom reluctantly became involved in atomic research. Neither government was convinced that the creation of an atomic bomb was possible, but neither was willing to risk being left out if indeed a bomb was developed. At the end of 1939, the scientific community in the United States convinced President Franklin Roosevelt to appoint the Uranium Committee to explore nuclear physics, and in April 1940, the British government agreed to the appointment of the MAUD Committee to look into the feasibility of an atomic bomb. The MAUD Report of mid-1941 argued persuasively that the production of an atomic bomb was possible. This report caused a flurry of excitement in the United Kingdom and the United States among those who were privy to its findings. In the United States, scientists pressured Roosevelt to seek some type of collaborative effort with the United Kingdom. The British, however, were not too receptive to the idea of working with the Americans. Not until 1942, when a British scientific team visited the United States and saw the enormous progress the Americans had made on their own, did the British agree to atomic collaboration. By then it was too late; the Americans were fast losing interest. This chapter will consider the events leading up to the Americans' change of heart.

On January 30, 1933, Paul von Hindenburg, President of Germany, reluctantly appointed Adolf Hitler chancellor of Germany. This event ushered in a dark period of systematic persecution of German and European Jews by Hitler's Nazi regime. On April 7, the Nazi government promulgated the Law for the Restoration of the Career Civil Service, and thousands of Jewish scholars and scientists were dismissed from German universities. "Aryan" Germans boycotted Jewish businesses and openly assaulted Jews on the streets as the police looked the other way. Many Jews fled Germany, seeking havens in other countries. Some 400 anti-Semitic laws and decrees followed this law. The exodus continued throughout the 1930s. In the end, Germany lost thousands of its best scholars, including a quarter of its scientists. Closely accompanying Hitler's domestic program of "Germany for the Germans" was his foreign policy of "Europe for the Germans." As the British and French practiced their policy of appeasement, Hitler embarked on a German expansion with the reoccupation of the Rhineland in 1936, the annexation of Austria in 1938, and the dismemberment of Czechoslovakia by March 1939. The invasion of Poland on September 1, 1939, and the subsequent outbreak of World War II put the pieces in place for the development of an atomic bomb. The Jewish emigrés would add a rich source of atomic expertise to that already existing in England, France, and the United States, and the war would provide the rationale for expending enormous financial resources on the development of a bomb.

However, despite the German "brain drain," the result of Hitler's domestic policy, one of the most remarkable breakthroughs in nuclear technology occurred in Germany: the discovery of nuclear fission by Hahn and Strassman. Before the end of January 1939, confirmation of an atomic split was repeated in numerous European and American laboratories.[1] Confirmation and additional contributions were made by, among others, the Frenchman Frédéric Joliot and his colleagues, the Austrian emigré Hans von Halban, and the Russian Lew Kowarski. This team would be the first to announce that the emission of new neutrons and a tremendous release of energy accompanied the splitting of the uranium nucleus.[2] Their findings were confirmed a week later by Enrico Fermi, an Italian scientist who had recently settled in New York, and Leo Szilard, a Hungarian physicist who had fled Germany to settle in the United States. Not to be left out, Niels Bohr, the Danish theoretical physicist, suggested that fission was much more likely to occur in the light isotope of uranium, uranium 235, which constitutes only 0.7 percent of natural uranium, than in the more abundant uranium 238, which constituted 99.3 percent.[3] Despite these advances, however, and the assumption among some scientists that a bomb was theoretically possible, no proof of its feasibility existed.[4]

The discovery of fission also contributed to a new development, outside the arena of the physicists and chemists, that was destined to change the course of atomic research. The year 1939 marked the official entry of the governments of Great Britain, France, and the United States into the uncharted field of atomic energy research. It is worth tracing the introduction of these pioneer nations to the field, since all three would make some contributions to early nuclear research. The U.S. entry, unlike that of the United Kingdom and, in particular, France, was faltering and hesitant. As early as March 1939, three prominent emigré scientists—Szilard; Eugene Wigner, another Jewish physicist of Hungarian origin; and the Italian Fermi, who had just received a Nobel Prize—attempted to warn the U.S. government "that uranium might be used as an immensely powerful explosive."[5] In 1935, at the age of thirty-seven, Szilard, sensing the future plight of Jews under the Nazis, fled Germany. He resigned his post at the University of Berlin and went to London. Even before the Germans split the atom, Szilard saw the bomb-making potential of the atom. He tried to warn the U.K. authorities and in February 1936 actually turned over to the British navy his patents on how to make an atom bomb. Unable to gain financial resources for his neutron research, or even access to a laboratory, he migrated to the United States, bringing with him the obsessive fear that the Germans would produce a bomb first. He spoke out strongly against the publication of atomic information by other scientists for fear that the Germans would gain access to it. Once he had convinced himself about any issue or cause, he drove others to exasperation in attempts to make them adopt it as well. During the war, his single-minded cause was to develop an atomic bomb before the Germans did. His persistence, high energy, and restlessness became legendary among his colleagues.

The forty-year-old Fermi had fled Italy in 1938 after winning the Nobel Prize for his work in search of the transuranic elements. His wife, Laura, was Jewish. He was an amiable man, generally regarded by his peers as a brilliant experimentalist and a great theorist. He met Szilard at Columbia University. On December 2, 1942, as part of the University of Chicago team, he, Szilard, and about forty others would conduct a plutonium experiment. The result would be the first self-sustaining nuclear chain reaction. Wigner, a theoretical physicist, was the third member of the group. Polite to a fault, shy, and always well dressed, he carried himself like a well-bred European gentleman. After leaving Hungary, he taught at Princeton. He too would become a part of the Chicago plutonium team.

The three scientists met the American physicist George Pegram in New York to plan their approach to the government. Wigner believed that the discovery was too serious for them to "assume responsibility for handling it."[6] Pegram, the dean of graduate faculties at Columbia, gave Fermi a letter of introduction to Admiral Stanford C. Hooper, technical assistant to the chief of naval operations, who, accompanied by two naval officers and two civilian scientists attached to the Naval Research Laboratory, listened to Fermi's one-hour presentation on neutron physics. They were not impressed but were willing to maintain contact.[7] This meeting marked the first direct link between the physicists of nuclear fission and the U.S. government.[8]

As the year progressed and fission research evolved, others recognized what had already become abundantly clear to the American-based group: uranium was the key to atomic research. Professor George Thomson, a Nobel laureate in physics who taught at Imperial College, London, and Professor Lawrence Bragg of the Cavendish Laboratory were quick to appreciate the importance of tracking down the available supplies of uranium. They believed that the British government should take steps to place all sources of uranium under its control where the Germans could not get their hands on them. They conveyed their concerns to Sir Henry Tizard, chairman of the Committee on the Scientific Survey of Air Defence and a physicist by training, who accepted their theoretical arguments but expressed doubts about their practical military application. Nevertheless, he agreed that he could not ignore the single chance in 100,000 that a bomb could be developed.[9] He approached the Treasury and the Foreign Office, and they discovered that the Union Minière du Haut-Katanga, a Belgian company, held a stockpile of uranium in Belgium. A significant number of British investors, including Lord Stonehaven, its vice president, held stocks in the company. Union Minière was engaged in radium mining in the Belgian colony of the Congo in Central Africa. The uranium by-product, quite useless as far as the company was concerned, was stockpiled in Belgium.

Lord Stonehaven was asked to arrange a meeting with Edgar Sengier, the chief administrator of the company. Sengier, an engineer by training, had been involved with the company in an administrative capacity since 1911 and had fully directed it since 1932.[10] The revelation that uranium had acquired a new value did not take Sengier by surprise. Two days earlier, in Brussels, Frédéric Joliot had already made that disclosure to him. In fact, Joliot had tendered a proposal, to which Sengier was receptive, for a joint exploitation of uranium.[11] However, any vision Sengier might have entertained of striking a lucrative deal with the British government was quickly dashed when he arrived in London. The two-week interval since the initial approaches to the Foreign Office and Treasury had cooled the enthusiasm of Sir Henry. He now believed that the immediate importance of uranium had been greatly exaggerated. Hence, he could not justify asking the government to put up the large sum needed to purchase all available stocks.[12] The Admiralty too had its doubts, since there was no rush by others over the next few months to start buying up uranium stocks. It commented that either "foreign nations have limited funds to gamble with or . . . they have decided that the possibility of developing an explosive of unprecedented power from uranium is so remote as to be negligible."[13] Despite this note of pessimism, the British government jumped, admittedly ankle deep, into the sea of nuclear research. In July 1939, Professor Thompson was commissioned to conduct further experiments and to report his findings to the Air Ministry. The outbreak of war did nothing to speed up these experiments.

Two months later, after the outbreak of World War II, Sir Henry was still expressing his doubts "that any practical form of bomb could be made with uranium."[14] His doubts were shared by, among others, Winston Churchill, who, on the eve of the war, had written the Secretary of State for Air, on the advice of Professor Frederick Lindemann, who was soon to become Churchill's atomic energy adviser as Lord Cherwell. He cautioned that the development of a bomb was not imminent since scientists would not have the needed technology to extract the fissionable U-235 from uranium for several years. Even with that technology, he argued, the explosive powers of a uranium bomb would probably be no greater than those of existing explosives.[15]

Sir John Anderson, the home secretary, who was to become deeply involved in the British atomic energy program, was entirely skeptical that a chain reaction could be produced and was certain that even if it were, it could not be controlled. Anderson, the son of a Scottish publisher, embraced all the trappings of the English upper class: cigars, billiards, golf, roses, pinstripe trousers, and a formal demeanor. He joined the Colonial Office in 1905, at the age of twenty-three, and climbed rapidly through its ranks. In 1920, the government sent him to put down a rebellion in Ireland, and in 1932, he was appointed governor of Bengal, again to put down a rebellion and restore law and order.[16] He accomplished his mission after five years and returned to the United Kingdom with a well-earned reputation for toughness. This image endeared him to Churchill, who was not one to trifle with the colonial "natives" or those who threatened the unity of the British Empire. No less an authority than Sir James Chadwick, Nobel laureate, and the discoverer of the uncharged neutron, expressed similar reservations about the atom bomb. At the end of October 1939, he conceded that under the right conditions the creation of a bomb might be possible, but he could not see how this bomb could be prepared without blowing up immediately.[17]

By July 1939, while the British and American governments were inching forward conservatively toward nuclear involvement, Joliot moved aggressively to involve the French government in uranium procurement. Hans von Halban, Lew Kowarski, and a newly added mathematician Francis Perrin, a professor at the Sorbonne, assisted him. The following narrative is important not only because of the light it throws on the progress of the French atomic program and the French government's involvement in it but because both Halban and Kowarski were destined to play major roles in the British and American atomic energy programs. Moreover, they, and Joliot, would find themselves at the center of major controversies that developed between the Americans and the British during and after the war. By April 1940, Joliot and his associates had pulled slightly ahead of the pack by describing the construction of an energy-producing machine that, in effect, foresaw the essentials of our current nuclear power stations and their characteristics.[18] This patent and two others were secretly registered between May 1 and May 4, 1939, in the name of the Centre National de la Recherche Scientifique (CNRS). CNRS was the government agency that paid the salary of Halban and Kowarski and that had already granted substantial research funds to the College de France, where Joliot taught. The four men donated 80 percent of any profits to CNRS for the advancement of scientific research. Nine days later, and three days after meeting the British representatives in London, Edgar Sengier met the director of CNRS and the Joliot team in Paris. A partnership agreement was hurriedly drawn up between Union Minière and CNRS—an unusual arrangement involving a private Belgian company and an organ of the French state. Basically, it was an agreement to develop jointly and exploit the patents held by CNRS through the use of uranium provided by Union Minière. The clauses of the agreement left open the possibility of explosive experiments. In addition to providing uranium, the company agreed to prepare equipment for experiments and to finance the experiments. The profits were to be divided equally for the most part. Bertrand Goldschmidt, himself a French scientist at the time, argues that if the war had not broken out, Joliot and his team would have had a good chance at being the first to construct an atomic pile, and Union Minière would have controlled 50 percent of the royalties. One can add that France would have also controlled the largest known source of uranium deposits.

Other developments tend to support Goldschmidt's opinion. On the eve of the war, Joliot's team specified the conditions under which a chain reaction might occur. The secondary neutrons that were emitted at high speed when the nucleus of the atom was split had to be slowed down before they could penetrate other nuclei to cause new fission. A substance was thus needed to slow down the neutrons without absorbing them. Ordinary water absorbed too many neutrons. The team settled on heavy water. Harold Urey, an American chemist, had already discovered in 1931 that hydrogen contained a second atom that he called heavy hydrogen. When heavy hydrogen was combined with oxygen, it formed water that was called heavy water. The only company in the world that produced heavy water was the Norwegian firm of Norsk Hydro-Elektrisk Kelstofaktieselskab. The team solicited the help of Prime Minister Edouard Daladier and Finance Minister Paul Reynaud, who sent a secret mission to Norway in late February 1940—less than six weeks before the Germans landed—assisted by three secret service agents, to arrange a loan of the heavy water. The mission acquired a sense of urgency when it was discovered that the Germans were trying to purchase the stockpile of 185.5 kilograms. The French struck a deal to borrow the substance, and after duping the Germans into believing they had loaded the heavy water on an Amsterdam flight, they transported it via Scotland to France. German fighter planes forced the Amsterdam flight to land in Hamburg, where a futile search was conducted.[19] For the next two months, the Joliot team prepared for its experiment, but before it could be conducted, the Germans broke through at Sedan on May 16. Frantically, the government made arrangements to ship the Joliot team and its twenty-six cans of heavy water to England. At the last minute, Joliot decided to remain behind, where he could be of service to the French cause, and Halban, Kowarski, and their families fled to England with the heavy water.[20]

Meanwhile, events had taken and were taking place across the Atlantic that would launch the American atomic energy program. While the Joliot team was deciding that heavy water presented the best means to slow the neutrons, Fermi and Szilard in the United States had arrived, independently, at the conclusion that carbon, or rather its mineral form, graphite, was the most effective agent. Leo Szilard called together his Hungarian connection of Eugene Wigner and Edward Teller to discuss his plans and concerns. First, he had no means of raising the $35,000 needed to conduct his graphite experiment, and second, the Belgians had to be warned "against selling any uranium to Germany."[21] On July 16, 1939, Wigner and Szilard solicited the help of Albert Einstein, who was vacationing on Long Island. To their surprise, he had not yet heard of the possibilities of producing a chain reaction; however, he was quick to see the security implications.

They drafted a letter to the Belgian ambassador, but before it could be delivered, another emigré, Gustav Stolper, a German economist, placed them in touch with Russian-born Dr. Alexander Sachs, a vice president of the Lehman Corporation and a friend of President Franklin Roosevelt. He scrapped the idea of contacting the Belgians and, to a stunned Szilard, suggested going straight to the White House. Einstein and Szilard redrafted a longer version of their letter, which they gave to Sachs on August 15 for transmission to the president.

President Roosevelt could trace his background to the "aristocracy" of New York State, where his relatives had lived for 300 years. His family estate was located in Hyde Park overlooking the Hudson River. In that mansion the young Roosevelt grew up rubbing shoulders with the privileged political insiders of the state and the Democratic Party. After graduating from Harvard, he became involved in state politics. By 1918, he had already earned an appointment in President Woodrow Wilson's administration as undersecretary of the navy. It was in that capacity that he first met Churchill, who addressed the Americans in London during the waning years of the war. Churchill, then minister of munitions, would have no recollection of this early introduction to the junior U.S. official. Roosevelt, however, remembered it because of Churchill's arrogance and indifference to the Americans' role in the war. In early 1921, Roosevelt contracted polio and lost the use of his legs, thereby interrupting his political career. But due to his personal resiliency and charisma, he resurrected his career to become governor of New York in 1929. Four years later, in the wake of the Wall Street crash and the subsequent depression, he was sworn in as president of the United States. His "New Deal" helped to revitalize the U.S. economy and pull it out of the worst stages of the Depression.

The president was midway through the second of his four terms in office when Einstein and Szilard sent him the uranium letter. The letter was not presented to him immediately because, among other things, he was preoccupied with the outbreak of the European war. Not until October 11, 1939, six weeks after the outbreak of war, did Sachs present the letter to Roosevelt. He gave a full oral explanation and made a recommendation that Roosevelt "designate an individual and a committee to serve as a liaison"[22] between the nuclear scientists and the government. Roosevelt agreed that action was required. An Advisory Committee on Uranium was quickly set up, consisting of Dr. Lyman Briggs, director of the Bureau of Standards, and two military representatives, Lieutenant Colonel Keith F. Adamson and Commander Gilbert C. Hoover, to work with the nuclear physicists. And thus was born the U.S. atomic bomb project.

The appointment of Briggs's Uranium Committee did not immediately translate into full-scale government action. The committee held its first meeting on October 21 and subsequently submitted a report to the president on November 1 that recommended government support for a full investigation of atomic energy. Roosevelt read the report and inexplicably filed it, where it languished until one of his staff members resurrected it in February 1940. The paltry sum of $6,000 in research funds was released. Meanwhile, even though the government was slow to act, fission research continued at many American laboratories. However, almost all of the men engaged in these studies regarded the development of a uranium bomb as "a remote possibility at best."[23]

Britain took the next major step in the bomb project. In March 1940, Otto Frisch and Rudolf Peierls, another German refugee, wrote a secret memorandum that pursued the line of thought first raised by Bohr. Bohr had stated that of the two uranium isotopes U-238 and U-235, the latter always experienced fission from neutron bombardment. The memo suggested that if U-235 could be separated from U-238, the pure U-235, subjected to fast-neutron bombardment, would eject high-speed neutrons during the fission of a single atom. These neutrons would produce an exponentially growing infinite number of fissions in a few kilograms of U-235. In this process there would be no need to slow down the neutrons to prevent their capture by U-238. Moreover, the bomb would not require the tons of uranium that scientist had predicted but only a few kilograms of U-235. The two men even suggested a detonation mechanism and raised the possibility of radioactive substances. Gowing describes the memorandum as "a remarkable example of scientific breadth and insight. It stands as the first memorandum in any country which foretold with scientific conviction the practical possibility of making a bomb and the horrors it would bring."[24]

The memorandum led to a flurry of activities in England. Peierls and Frisch showed it to Mark Oliphant, director of the physics department at the University of Birmingham. He sent it to Sir Henry Tizard, who in turn sent it to George Thomson, who brought in Professor John Cockcroft, a physicist who worked with the Ministry of Supply. The government quickly created a uranium subcommittee in April with Thomson as chairman. By June it was placed under the Ministry of Aircraft Production and given the unassuming code name "the MAUD Committee." For the next year the MAUD Committee considered the feasibility of an atomic bomb.[25]

Ironically, the two men whose work had served to inspire the formation of the committee were never given membership due to their refugee status. Nonetheless, they and a high proportion of European aliens were destined to make research contributions in the field. After the war and especially during the period of spy scandals, questions would arise surrounding the prevalence of foreign-born nationals in the British atomic program. Part of the explanation for their enlistment lay in the shortage of chemists and physicists during the inception of the program because many of the British scientists were already consigned to work in other areas of war-related research. In October 1940, the ministry was forced to restate its policy in the following terms: "We are in principle strongly opposed on security grounds to the employment of aliens on work of a secret nature unless it can be shown that every effort has been made to obtain British personnel and that the technical branch concerned is of the opinion that it is in the national interest to make use of their services."[26] Making use of their services could not be avoided, since many of them were affiliated with the same nongovernment institutions through which the committee worked. By mid-1940 the MAUD Committee had contracted out the research work to the universities of Cambridge, Oxford, Liverpool, and Birmingham and to Imperial Chemical Industries (ICI), a private research and development company.

Meanwhile, the Germans were moving through Norway and Denmark with breathtaking speed. The Labour Party, totally disillusioned over Chamberlain's handling of the defense of Scandinavia, engineered a revolt in the Commons that led to Chamberlain's removal from office. On May 10, Winston Churchill succeeded him as the new prime minister. Churchill began his political career as a member of the Tory Party. But in 1905 he "crossed the floor" to join the Liberal Party. During World War I the Liberals appointed him first lord of the Admiralty, the cabinet minister under whom the navy fell. After the Gallipoli disaster, an expedition that he had engineered against the Turks, he was dismissed on the insistence of his former Tory associates as a condition of their joining a coalition government. Before the war ended, David Lloyd George, the coalition prime minister, brought him back as minister of munitions. In the 1920s, he returned to the Tories, who shared his distrust of the Soviets, as chancellor of the Exchequer, the most influential of the cabinet appointments. He lost office again in 1929 when the Labour Party won the election. What followed were several years of financial distress caused by his losses in the stock market's failure. During the 1930s, Churchill had tried to alert the United Kingdom and the world to the danger of Hitler and the Nazis. German rearmament alarmed him, and he had appealed to the British government to build up its military forces and form an alliance with the Soviets. His colleagues, especially the appeasers within his party, dismissed his warnings as those of a warmonger. Moreover, due to his extreme imperialist views, and his habit of criticizing his Conservative Party's policies after it regained control of the government, he was dismissed to the backbenches of Parliament. It appeared that any aspirations for political power and influence were unrealistic. But when the war he had predicted came, his political career and popularity experienced a resurgence. The Tories turned to him by appointing him first lord of the Admiralty. President Roosevelt, probably sensing that Prime Minister Chamberlain's weeks were numbered, broke protocol by sending Churchill a personal, informal letter seeking to be kept informed of developments in London. Thus was born their great friendship. Roosevelt's initial letter set the tone for the informality and personal interjections that characterized many of their later correspondences. Churchill's accession to power, at the age of sixty-six, coincided with the long-awaited German invasion of France. The French quickly crumbled, and by June 24 France was out of the war. The Germans divided the country and allowed the aged Marshal Pétain to rule the south with a capital at Vichy, while Hitler occupied the north and west. A small "Free French" force managed to escape to England under General Charles de Gaulle to continue the struggle.

It was at this juncture that Halban and Kowarski arrived in London with their supply of heavy water to be welcomed by the British scientific community Although the committee was concentrated on fast neutrons and U-235 and the "Frenchmen" on slow neutrons and heavy water, they were invited to remain in England and not go to Canada or the United States, where they thought it would be safer to continue their research. Britain was far from safe. It experienced nightly bombing raids from the German Luftwaffe through 1940 and, on a smaller scale, into 1941. In accepting employment in the United Kingdom, Halban, speaking for himself and Kowarski, entered into an agreement with the British government transferring to it certain of the French patent rights and, on their part, receiving an agreement that the British would reassign certain patent rights to the French government. This agreement would cause some irritation in Anglo-American relations before the end of the war and will be discussed later.

Undoubtedly, by late 1940, the British interest in atomic research was at a higher level than that of the Americans. Both sides, however, saw the advantage of cooperation. In the early fall, Winston Churchill established the first real contact between American and British nuclear scientists since the outbreak of war. He sent over a scientific mission under Sir Henry Tizard, and including Cockcroft, to exchange scientific information relating to weapons and inventions. The British team discovered that the Americans were proceeding along parallel lines with the work being undertaken in Britain. However, despite all the American resources, "nearly all the work in America seemed to be several months behind that carried out in Britain."[27]

Both sides agreed that closer contacts should be maintained in the future, and to facilitate this the British, in the spring of 1941, opened a British Scientific Office attached to the British Supply Council in Washington. In their turn, the Americans established the Liaison Office of the recently formed National Defense Research Council (NDRC) in London. In June 1940, Roosevelt had established the NDRC and appointed as its director Dr. Vannevar Bush, who had suggested its creation to the president. Bush, an engineer, inventor, and applied mathematician, was an efficient administrator who held the post of president of the Carnegie Institution and had served as vice president of the Massachusetts Institute of Technology. He saw the NDRC as the means to get the scientific community to contribute toward military research. The NDRC immediately absorbed the Uranium Committee, giving it "an articulate lobby within the executive branch."[28] For the present, however, cooperation really began in the early months of 1941 with the opening of the two offices in London and Washington. During this phase the British gave more than they received because they had more to give.

James Conant, the forty-six-year-old president of Harvard and an associate of Bush, confessed that the first he had heard about even the remote possibility of a bomb was when he traveled to England to open the Liaison Office in March 1941. When one considers that he was also the chairman of the Chemistry and Explosive Division of the NDRC, his knowledge of current nuclear developments did not augur well for the future of atomic energy commitments. Conant was an eminent scholar and chemist. He came out of World War I with the rank of major for his contribution to the development of poison gas. After the war, he taught at Harvard, where he became chairman of the chemistry department in 1931. By then, he had already earned international acclaim for his work on chlorophyll and hemoglobin. In 1933, at the relatively tender age of forty, the boyish-looking Conant was nominated president of Harvard University. His election surprised friends and critics alike, who questioned his limited administrative background. He soon won over most of his critics with his quiet charm, evident administrative skills, decisiveness, and emphasis on scholarship. On June 14, 1940, Bush recruited him into NDRC to help mobilize American scientists. However, they were not convinced that their effort should be expended on uranium research. Ernest Lawrence, a Berkeley physicist working on isotope separation, drove himself to exasperation in his attempt to get Conant to "light a fire under the Briggs committee."[29] What it earned him was a threat from Bush to either toe the line or be excluded from the government's program.

The British continued to pursue diligently the possibility of constructing an atomic bomb. They were convinced that the experimental work on it had gone beyond the realm of speculation. No such clear-sighted conviction existed in the United States. In April, Kenneth Bainbridge, a Harvard nuclear physicist, went to England as part of the liaison team to work with the British. He was allowed to sit in on a MAUD meeting, where he found to his surprise that the committee had "a very good idea of the critical mass and [bomb] assembly [mechanism]."[30] He rushed off a report to Briggs, who contacted Bush. Confessing that he was no atomic scientist and that "most of this was over my head," Bush turned to the National Academy of Sciences for help. At the end of April the academy formed a review committee, headed by Arthur Compton, a Nobel laureate and physicist at the University of Chicago. Three weeks later it produced a report that stated basically that a bomb was possible but not before 1945. The report was more optimistic about the power-producing possibilities of uranium, a finding that disappointed both Bush and Conant, who were skeptical that uranium had any military significance. It also recommended six months of intensive study before any firm decisions were made. Meanwhile, Bush persuaded the president to agree to a reorganization of the NDRC. On June 28, he created a new agency with wider control over all government science and research, the Office of Scientific Research and Development (OSRD). Bush moved up to head the new organization, reporting directly to the president, and Conant was placed in charge of NDRC, reporting to Bush.

In early July another National Academy committee was appointed to review the report of the first. Lawrence submitted a report on the importance of plutonium and fast fission, but the committee ignored his report and strongly proposed a slow-neutron chain reaction for power production.[31] It downplayed the possibility of a bomb. By July 1941 Compton believed that the government was very close to ending research in fission. But that decision was not taken. Events in London forced the OSRD to take a harder look at atomic research.

On June 22, 1941, Germany invaded the Soviet Union on a front that stretched from the Baltic to the Black Sea. Three weeks later, on July 15, the MAUD Committee approved its final report. The first of its two reports on the utilization of uranium for an atomic bomb stated:

> We have now reached the conclusion that it will be possible to make an effective uranium bomb which, containing some 25 lbs. of active material, would be equivalent as regards destructive effect to 1,800 tons of TNT and would also release large quantities of radioactive substances, which would make places near to where the bomb exploded dangerous to human life for a long period. . . . The material for the first bomb could be ready by the end of 1943. . . . Even if the war should end before the bombs are ready the effort would not be wasted, except in the unlikely event of complete disarmament, since no nation would care to risk being caught without a weapon of such destructive capabilities. . . . It may be mentioned that the lines on which we are now working are such as would likely to suggest themselves to any capable physicist.[32]

The report recommended that collaboration with the United States be continued and extended in the field of experimental work. The United States would not receive an official copy of the report until October 3, but Charles C. Lauritsen, a Caltech physicist who was working out of the NDRC and who happened to be in London, was invited to sit in on the MAUD meeting when the report was presented. On his return to the United States a week later, he briefed Bush,[33] who still refused to act.

The following month, Mark Oliphant, the Australian-born director of the physics department at the University of Birmingham and an original member of the MAUD Committee, came to the United States to work with the NDRC on radar. It was Oliphant who discovered, to his amazement and distress, that Briggs had locked away the reports sent from England on the preliminary findings of MAUD. When he addressed Briggs's Uranium Committee, he was astounded by how little Briggs had told them. In desperation Oliphant made the rounds of the scientific community, also touching bases with Bush and Conant, bringing them up to date. Leo Szilard commented after the war: "If Congress knew the true story of the atomic energy project, I have no doubt but that it would create a special medal to be given to meddling foreigners for distinguished services, and Dr Oliphant would be the first to receive one."[34] Nonetheless, Bush did not move until Thomson forwarded an official copy of the report in October. Bush had a good reason to procrastinate. On July 16, he had informed the president that the OSRD believed that the possibility of developing a bomb in the foreseeable future was remote, so that the United States would not be justified in diverting scientists to such a project. He did recommend, however, a small inexpensive program, more to keep up with work being done in Europe.[35]

Meanwhile, the report was having its desired effect in England. The highest level of government had become involved and was faced with tough decisions. Two of the toughest concerned whether to proceed with a bomb project and whether the project should be based in Britain or the United States. In his report to Bush on the MAUD meeting that he had attended on July 2, Lauritsen stated that the committee discussed in great detail whether the work could best be done in Britain, the United States, or Canada. The consensus, according to Lauritsen, was that it would be difficult, if not impossible, to undertake the project in England "at this time." Memory of the Battle of Britain was fresh in everyone's mind; thus, the risk of aerial bombardment was real. Moreover, there was the question of cost and the availability of a sufficient number of scientists to work on the project. Lauritsen reported that he had spoken individually to eight of the twenty-four physicists present and that they all felt very strongly that the United States should undertake the project.[36]

Lord Cherwell, the former Frederick Lindemann, and now Churchill's atomic energy adviser, did not agree. In 1941, Churchill and Cherwell had already been close friends for twenty years. Cherwell was born in Germany to extremely wealthy English parents. He was a man who throughout his life had rubbed shoulders with the elite of European society, including the German kaiser, the Russian czar, and the king of the United Kingdom. He also had a lifelong interest in science and during the 1920s began teaching at Oxford, where he directed the Clarendon Laboratory. The prime minister relied upon him for all his atomic energy information. On August 27, Cherwell sent the prime minister a memo that summarized the MAUD Report and urged the creation of an atomic bomb program before the Germans developed a process ahead of the British. He admitted that there were certain obvious reasons against constructing a plant in England. However, he argued, the United Kingdom would be better able to maintain secrecy since they already had wartime security measures in place. But the most compelling reason for possessing the bomb, in his view, was the fact that whoever possessed it would be able to dictate terms to the rest of the world. He added: "However much I may trust my neighbor and depend on him, I am very much averse to putting myself completely at his mercy. I would, therefore, not press the Americans to undertake this work; I would just continue exchanging information and get into production over here without raising the question of whether they should do it or not."[37] In the same memo he recommended that Sir John Anderson, who had "researched uranium in early life," take charge of the project on a ministerial level. Churchill received the recommendations with a certain ambivalence. He sent a memo to General Hastings Ismay of the Chiefs of Staff Committee three days later in which he indicated that he was quite satisfied with the existing explosives; however, "I feel we must not stand in the path of improvement."[38] He endorsed Lord Cherwell's recommendation to proceed and accepted Sir John Anderson as the responsible cabinet minister. He then asked for the Chiefs of Staff recommendation. Three days later they came out "strongly in favor of its development."[39] The British atomic bomb project was on its way. Its code name was Directorate of Tube Alloys.

Across the Atlantic, Bush was still exercising caution. But what if the British were on to something? Bush decided to take precautions. In early August he and Conant called in Charles Darwin, head of the British Central Scientific Office in Washington, and asked him to convey to the British authorities American interest in a joint project to develop a bomb. Darwin forthwith submitted a handwritten report to Lord Hankey, paymaster general and chairman of the Scientific Advisory Committee. The Americans did not receive a reply to this proposal until March 1942. Meanwhile, Conant's doubts had begun to evaporate. During the course of the summer, first George Kistiakowsky, a Harvard colleague who led the NDRC division on explosives, and then Lawrence and Compton convinced Conant of the feasibility of a bomb.[40] Conant passed on their conviction to Bush, who was preparing to report to the president. On October 9, six days after official receipt of the MAUD Report, Bush brought the president abreast of developments in England. Accompanied by Vice President Henry Wallace, whom he had already briefed, Bush gave Roosevelt a summary of the British findings and discussed the likelihood that the Germans, with their first-rate scientists, might have come to the same conclusions as the British and were probably even more advanced. If the MAUD Report was accurate, and only time would tell, Bush was not willing to risk the United States' falling behind. He convinced the president that steps had to be taken to keep the United States abreast of the new developments. The president gave him the green light to explore the feasibility of a bomb.

Acting on Bush's recommendation, Roosevelt cabled Churchill two days later, suggesting a "jointly conducted" or coordinated effort.[41] This cable was not well received in London. The MAUD Committee had recommended continued cooperation with the United States, and some members had even suggested transferring the work to America. Initially, even Anderson had proposed an American location. However, from the outset Cherwell had other ideas. Cherwell, as he had divulged in his memo to Churchill, had already concluded that the possessor of the bomb would be able "to dictate terms to the rest of the world." He was not quite willing to accept a joint effort or a U.S. plant, especially since the consensus, even in American circles, was that the British were ahead of the United States. At the same time he could not risk alienating the Americans, whose support Britain needed in its fight against the Germans, by an outright rebuff. Until the British sorted out all the issues involved in collaboration, Anderson suggested that "the Prime Minister might send some interim reply to the President assuring him in general terms of our desire to collaborate."[42] Churchill's response had been two months in coming, and when it finally arrived it was still vague and noncommittal: "I need not assure you of our readiness to collaborate with the United States Administration in this matter."[43] Churchill added that he had arranged for his people to meet with Dr. Frederick Hovde, Bush's liaison representative in London.

Churchill's response was timely. The Japanese had just bombed Pearl Harbor a few days earlier, and it was clear to Churchill and his advisers that the Anglo-American alliance was about to be formalized. Churchill and Roosevelt, in August 1941, had already agreed to the principles of the Atlantic Charter. At Placentia Bay both leaders had met each other for the first time since their initial introduction in 1918. They spent many informal hours aboard the American cruiser *Augusta* and the British battleship *Prince of Wales*[44] getting to know each other and laying the foundation for their friendship. The Atlantic Charter outlined some of the elevated principles around which the American people and their allies mobilized. These included the right of self-determination, no territorial aggrandizement at the expense of others, and the right of all nations to share fully in the trade and the raw materials of the world. The document left no doubt that President Roosevelt abhorred the expansionist policies of Germany and Japan. Earlier, at the end of 1940, the president had referred to America as the "arsenal of democracy," and by March 1941, both Houses of Congress had endorsed his Lend-Lease policy. Pearl Harbor provided him with the military and moral rationalization for a U.S. intervention that he long had known to be inescapable. Churchill's practical assessment of the attack could be measured by the same standard as that of his appraisal of the German invasion of the Soviet Union on June 22. At that time the United Kingdom had found another fighting ally. In reference to a Soviet alliance, Churchill had commented that his purpose was to destroy Hitler. He told his private secretary that "if Hitler invaded Hell I would make at least a favorable reference to the Devil in the House of Commons."[45] In dealing with the United States, the British, by making one vague concession on atomic matters, would help to show their sincerity in working toward the common good.

Meanwhile, Bush had energized himself. At the end of October the National Academy of Sciences submitted to him a third report, which he had commissioned to double-check the results of MAUD. This report, drawn up by Arthur Compton, strongly endorsed the British findings. By a remarkable reversal of emphasis, slow-neutron graphite and plutonium research were omitted from the report.[46] A new dawn had broken, it seemed, bringing with it insightful revelations. Bush had the report in the president's hand on November 27. The president in turn appointed a top policy committee to act as his nuclear advisers. The committee consisted of Vice President Henry Wallace, Secretary of War Henry L. Stimson, General George C. Marshall, Bush, and Conant. Stimson would later state that the policy adopted and pursued by Roosevelt and his advisers was "to spare no effort in securing the earliest possible development of an atomic bomb."[47] The technical branch of the project was also reorganized. Dr. Briggs, as chairman of Section S-1, the uranium section, supervised the physics end of the work. He reported to Conant of NDRC, who, in turn, reported to Bush, chairman of the umbrella organization OSRD. The scientific work within S-1 was divided among three program chiefs: Ernest Lawrence, Harold Urey, and Arthur Compton.[48] The most notable committee, however, was the S-1 Section Executive Committee, which oversaw the entire atomic bomb project. Conant assumed the chairmanship of this group, which included Harold Urey, Ernest Lawrence, Lyman Briggs, Arthur Compton, and Edgar Murphree.

By the end of the year there was little doubt in Bush's mind that the Americans had everything to gain from a joint program. But the British resisted. Bush had received a report from Urey and George Pegram, a physicist and dean at Columbia University, whom he had sent to Britain to hold discussions on uranium and collaboration. Both men were impressed by the progress of British research. Concerning collaboration, however, they were forced to defend the United States against British charges that the Americans were interested only in the industrial application of atomic energy and not in its bomb potentialities[49] (an ironic charge, since within the next two years the Americans would be leveling the same accusation against them). In November, Bush instructed Frederick Hovde, his liaison officer in London, to raise again the issue of collaboration and a complete exchange of information. The presentation was made to Cherwell and Anderson, who was lord president of the Council, in addition to being the sole cabinet minister responsible for atomic energy matters. Anderson replied that the British government was just as anxious to collaborate but that it was disturbed about the possibility of leakage of information to the Germans, who, if they found out that the British or Americans were pursuing a bomb project, would accelerate their own efforts. In Britain, he explained, they had means of preserving secrecy that were not so readily available in the United States because the United States was not at war. He advised Hovde that as a preliminary step to cooperation the United States must reorganize its project to ensure a maximum degree of secrecy. Hovde requested concrete suggestions from the British so that the Americans could model their organization along British lines.[50]

There are a few ironies involved in this haughty approach taken by the British. First, it turned out that the British program was infested with Soviet spies. Second, the British were the first to complain when the tight security arrangements adopted by the United States deprived them of atomic information. And third, despite their concerns, it took the British two months to submit their own organizational blueprint to the United States. By then Bush already had his own plan in place. As a matter of fact, a description of Bush's organization was given to Darwin on December 23,[51] but either Anderson had not seen it by January 20, 1942, or he was not impressed by it. When he transmitted the British organization to the United States, he made no reference to Bush's; rather, acceptance of the British organization was presented as the basis for cooperation.[52]

Were the British merely stalling, or were they concerned about American security? Beyond a doubt the British were security conscious about atomic energy—and would remain so even up to the present. General administrative supervision of the work was entrusted to the Department of Scientific and Industrial Research. This department was viewed as a good "cover" for military work because in ordinary circumstances it was responsible for the development of industrial projects. The director of the entire project was Wallace A. Akers, research director of ICI—a commercial association that would cause problems with the United States in the future. The code name Tube Alloys was designed to create a connection with airplane radiators and tanks. To advise him, Akers was given a small technical committee that included Chadwick, Francis Simon, and Halban. Anderson too had his own Consultative Council to advise him on policy matters; it included Colonel Moore-Brabazon, the minister of aircraft production, Lord Hankey, and Lord Cherwell. It was Anderson's option to seek or not seek the advice of the council. Very few people outside of these groups, including Clement Attlee, the deputy prime minister, knew much of anything about Tube Alloys. When one considers the magnitude of the project and the potential effect of its success, not to mention that the country was at war with an enemy that had the wherewithal to develop a competing project, secrecy was an absolute necessity. Moreover, the British felt that the Americans were too open about nuclear matters. Their scientists, with the possible exception of Szilard, published freely in journals and newspapers, and most talked openly about their research endeavors. The British saw the need to muzzle this free expression of atomic matters. In time the Americans agreed.

At the same time one cannot deny totally the stalling motive. Not everyone involved was convinced of the need for a joint project. In any joint project most of the plants would have to be built in the United States—with British expertise, one may add—which had the needed financial resources and was far removed from German bombing. The British were not willing to concede all the experimental work and plants to the United States. They believed that each country should pursue its own research and experimental work but that information should be exchanged freely. Halban's boiler project was the exception. At the end of 1941 the Technical Committee had recommended its transfer to the United States, where it could take advantage of the better facilities; however, the committee insisted that it remain under British control and not that of the United States.[53] The British, it seems, needed time to weigh options and decide on a policy that would serve British national interest. For the first six months of 1942, this policy of collaboration was pursued not tentatively as a joint effort but as two independent projects sharing research information.

Before the end of the summer of 1942, both nations experienced a change of outlook regarding collaboration. The British wanted more, and the Americans wanted less. How did this happen? Were the British slow to detect the astounding progress that the United States had made in the field of research during the first few months after the issuance of the MAUD Report? And did they fail to attempt to take advantage of this progress? As early as February 1942, a British scientific mission under the direction of Wallace Akers, director of Tube Alloys, on a visit to the United States was shocked at the rapid progress made by the United States. Akers, Simon, Halban, and Peierls traveled the length and breath of the country, freely exchanging information on a wide range of nuclear matters. The same complete freedom of investigation was allowed as had been accorded Pegram and Urey in England. They were invited to the meetings of the S-1 Committee and visited the major universities engaged in research work. What they saw and heard left them impressed with the state of the U.S. program. American advances were measured by the number of top-rank physicists and chemists working on atomic energy, the financial resources placed at their disposal, and the high level of organization and coordination.[54] Gowing states that "it was clear even by the late spring of 1942 that the Americans who had been well behind the British in the race for a bomb had drawn level and were indeed passing them by."[55]

John Anderson too was impressed with their report of American progress and commitment, but still he was not prepared to commit British resources to a joint effort. He finally decided on a belated response to Bush's inquiry about a joint project, transmitted through Darwin the previous summer. He apologized for the delay and explained that the policy committee was pleased with the existing exchange of scientific information. However, since the technical situation was relatively fluid, the committee thought it was inadvisable to decide on any detailed plans for full-scale development at present. They should wait until their intensive scientific work provided more information on which such a decision could be taken—perhaps in a few months.[56] On April 20, Bush replied that he agreed that they should wait until the pilot plants were in operation. "At that stage I feel strongly that it is highly desirable that future action should be considered jointly, and I will look forward to more explicit interchange of our plans when we arrive at that stage of progress."[57] To no avail Akers and his team pressed for an immediate combined Anglo-American effort. It appears that Bush still saw the merit of a combined effort.

When Michael Perrin, the secretary-general of Tube Alloys, visited four months later, he found that since Akers's visit the United States had made even greater progress. He wrote Akers that the Americans would very soon "completely outstrip us in ideas, research and application of nuclear energy and that then, quite rightly, they will see no reason for our butting in."[58] He thought that the time available for coordinating the American and British plans could be measured in less than a month. Little did he know the extent of his prescience. However, Anderson's Consultative Council was not in agreement. Lord Cherwell, in particular, still believed that it was possible for the British to construct a full-scale U-235 separation plant in England despite the financial and manpower demands of the war. In June, the council demanded a comprehensive report on the pros and cons of the project. But events were outstripping British procrastination.

By the summer none other than Anderson himself came around to accept the growing reality of the British program's position vis-à-vis that of the United States. He won over the council and sent the prime minister a two-page memo on July 30. He explained that it had become clear to him that the demands of the war made it impossible to erect a production plant in England, "even the erection and operation of a pilot plant would cause a major dislocation in war production." He confessed that the Americans were making rapid progress and that the British could not rival their "enthusiasm" and "lavish expenditure." Consequently, "with some reluctance," he was forced to conclude "that we must now make up our minds that the full scale plant for production . . . can only be erected in the United States and that, consequently, the pilot plant also will have to be designed and erected there. . . . We would move our design work and the personnel concerned to the United States. Henceforth, work on the bomb project would be pursued as a combined Anglo-American effort."[59] Moreover, he added, the British must face the fact that the pioneering work done in England had become a dwindling asset. But some benefits could still be salvaged, since the scientists who would be transferred to the United States "would keep abreast of developments over the whole field" and after the war "would enable us to take up the work again, not where we left off, but where the combined effort had by then brought it."[60] Anderson's rationale was sound and practical, but it was based on an unknown premise: that the American proposal for a joint program, tendered in 1941, was still on the table. In mid-1942, it was not. The Americans were not as anxious as earlier to pursue a combined program. They were fully aware of their own rapid progress.

Goldschmidt believes that the evolution of Anglo-American cooperation can be traced to the four men responsible for atomic energy policy making: Cherwell and Anderson on the British side and Bush and Conant on the American side. In reference to all four men, Goldschmidt states: "In this ballet of Anglo-American nuclear relations, the Americans Bush and Conant revealed themselves to be by far the best and most perspicacious advisers because they were less political and closer to the technology. Sensitive to British advances in the summer of 1941, they recognized the moment when the British began to lose ground. By contrast, the Englishmen Cherwell and Anderson, imbued with the antiquated dogma of imperial superiority, did not realize this reversal until it was too late."[61] Less than a year later, Goldschmidt adds, Bush and Conant "were to become and remain convinced opponents of the intimate collaboration . . . they had originally advocated. The ill-advised British government thus let the unexpected chance represented by the US offer [of a joint program] slip through its fingers."[62] Would subsequent events have unfolded differently if the offer of a joint program had been accepted at the end of 1941? It is doubtful. As we will soon see, the chain of events leading to the completion of the bomb and the issues contributing to the breakdown of postwar collaboration would have changed little. The experimental phase of the project would still have been conducted in the United States; British scientists still would have flocked to the United States to work on the project, and the United States still would have been the only country possessing the bomb at the conclusion of the war. The domestic and international problems raised by virtue of the fact that the United States was the sole possessor of the bomb still would have cropped up, thereby adversely affecting postwar collaboration.

In *The Hinge of Fate,* Churchill made a fascinating disclosure that during his visit to Washington in June 1942 to meet Roosevelt, he and the president had clearly agreed to a joint program.[63] If this was so, it would not be the last time that both men would make impromptu agreements without consulting their advisers or informing them of the decisions made. There is no evidence to indicate that Anderson or the others were informed of the discussions. We know that Cherwell was opposed to a combined effort and that Bush and Conant had already begun to back off from it. The signs of fading interest were there to be interpreted but were not recognized by the British. The next chapter will look at these signs. It will examine also the roller-coaster course of collaboration as it broke down completely and the British struggle to renew it.

TWO

Roosevelt Agrees to Atomic Collaboration: 1943

The bombing of Pearl Harbor exploded the myth that the United States could remain isolated from and untouched by the wars in Europe and Asia. The destruction of the American naval base in Hawaii instantly and inevitably produced a consensus for war against Japan and, after the German declaration, for war against Germany, which Roosevelt had long sought. Churchill could feel relief and satisfaction in January 1942 because Britain was no longer the sole Western bastion against Nazism. The Soviets were already relieving some of Britain's aerial bombardment by absorbing the brunt of the Nazi arsenal in the east, where the very survival of communism was threatened. And the Americans were about to hurl the might of their industrial and military power into the arena. The Grand Alliance had taken shape. By the fall of 1942 the German and Japanese momentum was checked on every front. The Allies counterattacked and by the end of 1942 were advancing in North Africa, holding their own against the Japanese, beating back the Germans on the eastern front, and clearing the sea lanes of the Atlantic. Caught up in the euphoria of battlefield cooperation and successes, the American and British leaders, if one accepts Churchill's claim, agreed to include atomic cooperation within the realm of their wartime alliance. But influential forces within the American atomic circle had other ideas.

In the period from mid-1942 until the end of the year, the United States discovered its increasing proficiency in the scientific and technical field of atomic energy. Bush and Conant took a second look at the combined effort they had once favored and decided that an independent American program best served the national interest. They recognized that the time was fast approaching, if it had not already arrived, when they would have more to give to than receive from the British. As the months sped by, it became clear to them not only that a bomb was possible but that it was realistically probable before the war ended and that it could be developed without, or with minimal, British help. Slowly they began to question the need, not only for a combined effort, but also for collaboration itself. As the Americans moved inexorably away from collaboration, the British used every means at their disposal to regain American acceptance. Those means included having their prime minister grovel at the feet of the American president and literally beg for a resumption of collaboration. This episode marked a low point in the dignity of the office of the prime ministry. It is doubtful whether the British ever regained the respect of American atomic energy policy makers or recovered their self-confidence in atomic energy research after this period of supplication.

The discussions surrounding the location of the Halban heavy water experimental plant presented the first bit of evidence that something was amiss. This pilot boiler, as it was called, was designed to produce plutonium from U-238. Both British and American scientists generally accepted that plutonium's military application was not yet as immediate or effective as that of U-235. It was viewed as a potential source of industrial power. Hence the emphasis in U.K. circles, thanks to the finding of the MAUD Committee, was on a diffusion plant to produce U-235. On the other hand, the Americans, as the British soon discovered, were engaged in four different methods:

1. *Gaseous diffusion.* Harold Urey, who won a Nobel Prize for his discovery of heavy water, led a team at Columbia University that was using this method in an attempt to separate U-235 from U-238.
2. *Centrifuge.* Edgar Murphree, an engineer with Standard Oil, worked out of Pittsburgh with a team engaged in a process that used a centrifuge to separate the U-235.
3. *Electromagnetic.* This method was used by Ernest Lawrence, the inventor of the cyclotron and another Nobel laureate, and his team at the University of California to separate U-235.
4. *Graphite pile.* At the University of Chicago, Arthur Compton, who had won a Nobel Prize for his work on X rays, headed an illustrious team that included Szilard, Fermi, and Walter Zinn. They constructed the first nuclear reactor for the production of plutonium.

In the interest of efficiency, both countries had discussed, in February and March, the desirability of centralizing the plutonium work by transferring Halban's heavy water team to Chicago to work in parallel with the American graphite pile team. As the British saw it, Halban's would be a completely independent British team, paid for by the U.K. government. The American Technical Committee and the Chicago scientists, including Compton, welcomed the idea of having the Halban team in close proximity. They were impressed with Halban's knowledge of heavy water, and although they were working w_th graphite they saw the advantages of cooperation.

Nevertheless, Bush shot down the idea with the argument that it would be impossible to get the FBI to grant clearance to the many non-British-born foreign nationals on the Halban team, most of whom were of German and French nationality.[1] Bush was also concerned about the independence of the British team. On May 13, he confessed to Wallace Akers, the leader of the U.K. team working on gaseous diffusion, that he could not support a U.K. team reporting only to the British government. It was too much of a "break with tradition."[2] Halban, on the other hand, firmly believed that Bush and Conant were against having the U.K. team in Chicago because its presence wc uld create contacts between both teams that would break the "narrow ver-. tical American secrecy organization."[3] Whatever the reason or reasons, the team was unwanted in America. The British were forced to consider a third alt ernative: transferring the thirty-man team to Canada.[4] On September 2, 1942, Halban was sent to Montreal to set up a heavy water reactor for the production of plutonium. Bush was happy with the new arrangement.

Despite Bush's successful opposition to the Halban team, the British still believed that the decision to pursue or not to pursue a joint program was theirs to make. The British, faced with the German menace, the demands on their scientific manpower, and the financial exigencies of the war, were beginning to lose confidence in their ability to complete a bomb project. On July 31, 1942, Churchill accepted the Consultative Council's recommendation for a fusion of the British and American projects, and Anderson sought Bush's acquiescence on August 5. In one of two letters (the second suggested the formation of a Joint Nuclear Commission and the working out of a patent agreement), Anderson explained that the British would be unable to complete their diffusion plant within two years and that hence he and the prime minister had concluded that the plant should be built in the United States. The United States would be responsible for arranging the design and construction of the plant, which would be added to the U.S. program. The British would then transfer Professors Simon, Peierls, and members of their research teams and other key men to work alongside the Americans. He indicated also that he expected that Bush "would be willing to add British members to Dr Conant's Executive Committee."[5]

I_ took Bush almost four weeks to decide how to extricate himself from the joint project that Anderson's letter seemed to take for granted. He called in Conant, and they sought input from Conant's Executive Committee. The result was a noncommittal response. The United States too, he explained, anticipated interference from the war effort and could not commit itself to the construction of American pilot plants in the future until each plant was evaluated and assigned a priority status in the entire war program. Until the American plants were evaluated and prioritized, the United States could not make foreign commitments. As for letting U.K. representatives sit on Conant's Executive Committee, that decision had to "wait further determination of the way and extent of integration of the two programs."[6] No promises were made other than to explore the matter further. In addressing Anderson's second letter, Bush stated that he was in favor of a patent arrangement and that he would explore further the proposal for a Joint Nuclear Energy Commission. Conant was quite impressed with both replies. In a handwritten note to Bush, he wrote: "Both letters are masterpieces! You should have been a lawyer or a Diplomat."[7]

If the "diplomatic" language of the September 1 reply still left doubts in the mind of Anderson concerning American interest in a joint project, Bush's follow-up memo of October 1 removed them. Bush emphatically turned down the idea of building a British diffusion plant in the United States. This was not possible, he explained to Anderson, because the Americans had reorganized their atomic program, placing it under the general supervision of the army. Indeed, the army had by then assumed control of the project. As early as March 1942, the president had instructed Bush to turn over the developmental stage of the bomb to the army. In late September, Bush completed the transfer from Office of Scientific Research and Development to the army's Manhattan Engineering District. Brigadier General Leslie Groves of the Army Corps of Engineers assumed direct control over the daily supervision of the project. Groves was the son of a minister who had inculcated in him a keen sense of patriotic duty. In the army, Groves had earned a reputation for tactlessness that bordered on rudeness, insensitivity, and downright boorishness. But he was also a tough, no-nonsense man who always got the job done. He was disappointed with his new assignment, even though it came with a promotion to brigadier general. He had just finished supervising the construction of the Pentagon and had hoped for a more glamorous overseas appointment, one that might advance his career. But it did not take him long to grasp the magnitude of the atomic project and to recognize that its success could become the defining achievement of his military career. He would allow nothing to stand in the way of that success.

Bush also created what he described as a sort of board of directors, a Military Policy Committee, with general supervisory responsibilities over the entire atomic energy program. He reserved for himself the chairmanship of this committee and included Conant, a representative from the army, and a representative from the navy. Bush described the new organization to Anderson. Conant's Executive Committee remained in charge of the scientific program, and Bush expected the U.K. scientists to continue dealing with it. He explained, however, that the new organization had decided to go ahead with pilot plants for the diffusion, centrifuge, and electromagnetic methods and also with the experimental graphite pile. In addition, it planned to become more involved in the heavy water experiments at Trail. This major commitment, Bush continued, would fully exploit the constructional resources of the United States. As such, the United States could not consider further the idea of building a British diffusion plant in the United States.[8] Both men agreed to send Akers over for a full briefing.

As soon as Bush and Conant received confirmation of Akers's intended visit, they began planning his reception. Conant was of the opinion that they should put their cards squarely on the table and point out the difficulties raised by Anderson's proposal. He strongly favored telling Akers that the United States was moving in the direction of guarding its information more closely than it had in the past. He thought that it was "clearly quite unfair to American scientists to compartmentalize so definitely this project when we let British scientists roam around the country picking up information from each one of the various projects."[9] Bush and Conant took the matter to Secretary of War Stimson. Three days later he told the president that "Bush and the others" were anxious to know what foreign commitments he had made concerning atomic energy. Roosevelt assured him that he had had discussions of a very general nature with no one other than Churchill. Stimson wrote that he had then asked the president "if it wasn't better for us to go along for the present without sharing anything more than we could help. He said yes."[10] However, Roosevelt suggested that they should hold a conference with Churchill to discuss the matter.[11] Churchill would initiate this discussion long after the new policy was adopted.

For the moment, though, on the basis of the argument Stimson had presented, it was clear to the president that Stimson believed that since the United States was "doing nine-tenths of the work," it was entitled to nine tenths of the benefits. Roosevelt has since earned a notorious reputation for saying what he believed his listeners wanted to hear and not revealing his hand on contentious issues or issues that were close to the heart of those with whom he conferred. We will see other glaring examples of this well-earned reputation in his future handling of atomic energy. Either way, his aides believed that they had convinced him that the United States should be more guarded with its secrets and that the United States could proceed without British help. Whether it was his intention or not, Roosevelt had given his representatives the approval they had sought to draw up a comprehensive policy for limited cooperation. By his evasiveness he had set in motion the events leading to the first stage in the long history of atomic noncollaboration.

Akers arrived in the United States in the first week of November 1942 and departed at the end of January 1943. When he arrived, the new policy had already begun to take shape; when he left, it was firmly in place. On November 4, he held a four-hour discussion with Bush and Conant during which the new organization was explained to him in detail, with special emphasis placed upon compartmentalization and other steps taken by the army to increase security. In the future, they explained, information would be exchanged on a "need to know" basis, a security device that both governments had used in the past. Extreme discretion would be exercised to prevent the transmission of information from one group to another if the other was not engaged in similar work. The special object of this approach was to prevent the leakage to Germany of any atomic energy information or any information that might expose the U.S. plants to sabotage. To observe this secrecy measure, the various working groups were instructed to commit as little as possible to writing. What information the United Kingdom received would be through a high-ranking British liaison officer, who would have to assimilate the information through verbal transmission. He would be the sole means of communication between the United States and the United Kingdom. Moreover, visitors from England would have to get official clearance from the U.S. authorities for each visit they proposed to make. On arrival in the United States, they would have to obtain written permission from Conant to visit the respective sites. To receive a written document from an American group, a U.K. visitor had first to make application through Conant's Executive Committee.

At the end of the discussions, Akers wrote that it was clear to him that Bush and Conant were far from happy about the restrictive attitude of the army.[12] His observation could not have been further from the truth. We will come across several instances where both men, through a plea of ignorance or outright duplicity, would attempt, sometimes quite successfully, to evade responsibility for or deny participating in the formulation of policies that offended the British.

On November 13, Akers saw General Groves, who confirmed the remarks about secrecy and gave the same reasons. On taking over the project, Groves explained, he had been horrified at the number of people who knew, or thought they ought to know, about large fields of the project beyond their particular area. He stated that his intention was to divide the work into as many separate compartments as could be devised and to fill each of the compartments with as many people as could possibly be used efficiently in them.

Only one person in each cell would be able to see over the top, he explained, and that person would not be able to see into more than a minimum number of other cells. The group leaders, members of the Executive Committee, would be the only people with a general knowledge of the overall project. As far as the British were concerned, they would be able to talk to the heads of all the S-1 groups, but discussions were to be confined to matters of common interest.[13]

Akers accepted the new arrangement with no protest. He wrote to Perrin: "I do not think that there is going to be difficulty over this, because the group leaders, such as Compton, Urey and Lawrence, will certainly be as helpful as they can."[14] Apparently, Akers did not fully appreciate that the secrecy policy was directed not exclusively against the Germans but equally against the British. So far, the Americans had taken pains to avoid the charge of Anglophobia. Until there was a comprehensive policy to lay before the president, they were willing to hedge a little while fine-tuning their recommendations. Also, Akers could not have foreseen the length to which Groves was prepared to go to make his system as airtight as possible. People like Lawrence, Urey, and Compton would become extremely cautious as Groves's security net fell over everyone remotely associated with the Manhattan Project. For his part, Akers did not believe the system would work. He thought that the British would have a short time to wait before the theoretical scientists rebelled against the constraints of the system.

On December 15, in the midst of Akers's lobbying effort for fuller collaboration, Bush, Conant, and Groves submitted their comprehensive policy to the Military Policy Committee. Although they were not members of the committee, the vice president and the secretary of war participated in the discussions. The other member was the Chief of Staff General George C. Marshall—Bush was its chairman.[15] On that same day, Akers submitted to Conant a long letter, which Bush, Conant, and Groves had requested some time earlier, outlining the British position on collaboration.[16] Akers had not yet seen the policy report; hence, he based his arguments on issues that had been raised in earlier conversations. It is worth comparing both positions.

First, regarding work on electromagnetic method, the U.S. position was that no further information should be given to the British or Canadians, since neither one was working on this method. Akers argued that information should be shared because the progress of any one method had direct bearing on the program under consideration for other methods. His second argument, a political one, was that it had always been the intention at the highest level that the development of the atomic project should be a cooperative effort between the two countries, regardless of the origin of the process used.

Second, regarding the diffusion plant, the Americans believed that exchanges should be restricted to design and construction information between U.S. firms and their counterpart in Britain working on the same project. This argument was based on the fact that the British were further along in the development stage of a diffusion process and hence had helpful information to provide in the construction of an American plant. The British stand was that collaboration should be extended to include the erection and operation of plants. Moreover, U.K. scientists and engineers should be actually employed in the operation of the plants.

Third, regarding the manufacture of heavy water and the production of plutonium, no further information was to be transmitted to the British or Canadians on the production of heavy water in the United States. Since the Canadian-British group would be working on the scientific branches of chain reaction, exchanges would be provided between the group at Montreal and that at Chicago. However, no information would be provided on the properties or production of plutonium, including methods of extraction or design of power plants. The British argued that their engineers should participate in design, construction, and operation of any full-scale plants in the United States. Furthermore, the United States should confine itself to the graphite process, as per the earlier agreement—which Conant disputed—while all heavy water experimentation should be left to the British and Canadians.

Fourth, regarding fast-neutron reaction and all matters concerned with the bomb, no further information would be shared with the British and Canadians. The British wanted full collaboration in this area. Akers objected to the isolation of a portion of the American scientific group, even if interchange with those scientists outside of isolation continued, on the grounds of inefficiency.

Finally, the Americans agreed to share basic scientific information only on direct approval of the chairman of the S-1 Executive Committee.

Ironically, the United States used Anderson's letter of August 5 as the basis to justify noninterchange of information. The report to the Military Committee made copious references to the letter in order to illustrate that the United Kingdom, by its own admission, did not have adequate facilities for the construction of the plants needed for the manufacture of U-235 and plutonium. "They have no intention," the report stated, "of engaging in manufacture in this war. Therefore, our passing our knowledge to them will not assist the British in any way in the present war effort."[17] The bottom line, however, was that the United States did not believe they needed the assistance of the British to construct a bomb. The report admitted that the British diffusion process was more developed but surmised that a complete cessation of exchanges in that field would hinder "but not seriously embarrass the United States effort." Similarly, it conceded that it would be advantageous to utilize the talent and knowledge of Halban's group, "but it would not hamper the effort greatly if the cessation of interchange resulted in the withdrawal of this group from the effort."[18] Would it be unfair to the British if the Americans ended collaboration? No, the report argued, because the British had made no greater contribution to the field of knowledge than the Americans had made.[19]

The Military Policy Committee approved the report, and it was sent, with Akers's letter attached, to the president, who in turn gave it his stamp of approval on December 28. Two weeks later, he left for Casablanca to hold a summit with Churchill. Why he made such a momentous decision on the eve of the summit is unknown. It is likely that he had not yet fully grasped how intimately involved Churchill had become with atomic energy. Lord Cherwell took quite seriously his responsibility for keeping his prime minister fully informed about atomic energy. If the president had known, he might have postponed his decision, thereby avoiding a potentially confrontational situation. A successful summit was important to the president. The midterm congressional elections had depressed him; his Democratic Party lost forty-four seats in the House and nine in the Senate. The Republicans had capitalized on rising inflation and on problems surrounding the mobilization effort. Moreover, after their successes in North Africa, it was time for him and Churchill—Stalin had turned down their invitation to attend—to plan their next major move against the Germans.

By a convenient coincidence, the atomic report arrived on the president's desk at almost the same time as a copy of the Anglo-Russian Agreement for the exchange of scientific information. This agreement had been signed the previous September; however, according to Margaret Gowing, this was the first time the president or Stimson had heard of it. She adds that the agreement had been made with the complete knowledge and approval of the U.S. government, which had approved the list of items to be exchanged. Gowing states that "it seems that the discovery [of the agreement] impelled the President to agree to restricted interchange on atomic energy."[20] The implication is that the president was concerned that the British might exchange atomic information with the Soviets. One doubts very much that the president based this momentous decision on the fortuitous appearance of this agreement on his desk. More likely, he was swayed by Bush's argument, especially the speculation that the United States could start up the production of a bomb by the first half of 1945, without outside help. Moreover, we should not forget that he had initiated the policy of noncollaboration by giving his aides his tacit approval of this policy on October 29.

Martin Sherwin does not believe that Bush and Conant recommended this restrictive form of exchange because the U.S. atomic program had been transferred from civilian to military control, because the United States was doing all the developmental work, or because of a concern for security—although each is a strong reason in itself. Rather, he argues, Conant believed that continued cooperation would jeopardize U.S. postwar commercial interest. "Conant perceived the British as displaying an inordinate interest in 'information which appeared to be of value to them solely for post-war industrial possibilities.'"[21] There is no doubt that industrial interest played a role in the minds of Conant and others. However, the overriding consideration was the recognition that the bomb could present them with a formidable weapon that would elevate their military arsenal and power to heights never before achieved by any nation. They were not willing to share this power with any nation, especially not the British, whose decline would provide a vacuum that the Americans were all too willing to fill. As Hershberg so succinctly puts it, Conant equated "the prospective weapons with the attributes of global leadership[;] he disdained to help Britain cling to American coattails as it was edged aside."[22] Conant's and Bush's postwar perception of atomic weapons will be further discussed below.

Akers was given an official copy of the new guidelines for restricted exchanges on January 13, 1943 (it was dated January 7), which he sent to London. Anderson had already seen the Canadian copy a few days earlier. He was shocked when he read it; he had been clinging to the hope that British differences with the Americans could be resolved. He wrote Churchill on January 11 that "this development has come as a bombshell, and is quite intolerable." He was convinced that Roosevelt was unaware of the decision to end collaboration because "it cannot be his intention that we should be treated in this way." Therefore, he suggested that the prime minister raise the issue with the president and urge him to resolve the matter.[23]

One should add also that British officials were also of the opinion that Bush and Conant had been forced to go along with this new policy. Bush and Conant took pains to convey this impression. In informing the Canadians of the new policy on January 2, Conant wrote that he and Bush had received the order from the top.[24] In a letter from the office of the British high commissioner in Canada to Anderson, discussing the uncooperative attitude of the Americans, the writer stated: "I have reason to believe that Conant himself disliked writing the letter, which probably does not express his own sentiments or those of the scientific community, but seems to be an expression of the somewhat high-handed and exclusive policy being adopted by some senior officers of the United States Army."[25]

It is true that as Groves tightened his control over the many facets of the Manhattan Project, Bush and Conant were shoved to the side. In the Military

Policy Committee, they were called upon more and more to express their views on general policy matters and less and less to express them on specific scientific matters. However, it is doubtful that they were as naive about the scientific side of the matter as they led Akers to believe. Nonetheless, the decision of December, which both men played the leading role in formulating, was a policy decision, centered on whether the United States should collaborate with the United Kingdom, and not a scientific decision. Conant and Bush conveniently appealed to their scientific shortcomings and nonassociation with the scientific side of the project to absolve themselves of responsibility for the new policy. The truth is that in December they were still both intimately involved in the scientific and policy decision-making process. Akers, however, continued to understate their role. On January 30, 1943, Akers commented to Dean Mackenzie, president of the National Research Council in Canada, that Conant had "practically dropped out of this business. He certainly no longer has any authority . . . [and] merely acts as a 'post-box.'"[26] Not exactly.[27] He and Bush were still called upon to express their views on policy matters.

It is worth emphasizing, however, that Groves and the military were not responsible for introducing the idea of noncollaboration. This idea, as the preceding discussion has shown, was already in the works before the army assumed full control of the bomb project. Bush and Conant passed the sword to General Groves, who in turn resisted collaboration with a vehement passion that both men never came close to matching. In fact, both men would not resist compartmentalization and "need to know" either. As they saw it, a tight security system was required to prevent leaks to the Nazis. They and other scientists believed that the Germans too were pursuing the development of an atomic bomb. On May 14, 1942, Conant had warned Bush that the Germans "may be already ahead of us by as much as a year."[28] Seven months later, on December 16, Bush informed the president that he still did not know "where we stand in the race with the enemy toward a usable result, but it is quite possible that Germany is ahead of us and may be able to produce superbombs sooner than we can."[29] In truth, the Germans had already relegated atomic research to a secondary priority to other secret weapons projects.[30]

The British wanted an explanation for the new policy. They could not understand the reasons behind the complete about-face of the Americans. They were not satisfied with the original explanation given by Groves and Bush of secrecy to prevent leaks to the Germans. Bush hinted to Akers, still in the United States, on January 18, that the "directive from the top" was connected with the recent Anglo-Russian Agreement (though one should keep in mind that the report was already drawn up when it met the agreement by chance on the president's desk). He explained that the United Kingdom had entered into the agreement without warning the United States that it was under consideration. He requested Akers not to quote him as the source of the information.[31] Akers was able to ascertain from London that the United States had been kept informed of the agreement from its inception and that the U.S. ambassador in London had discussed it with the appropriate officials in the United States. The Anglo-Russian Agreement was a red herring thrown out by Bush. Dean Mackenzie, the Canadian, suggested that Conant might be more forthright with a Canadian, so he offered to discover the source and reason for the directive. After meeting with Bush and Conant, he returned to Canada feeling "reasonably satisfied that he has got to the bottom of American motives." The directive, he reported, had come from the Military Policy Committee and was based on postwar security considerations. The Americans were concerned that the numerous foreigners on the British team would acquire secret information that they would pass on to their respective governments.[32] Postwar security was a concern, but certainly it had not originated in the Military Policy Committee.

Meanwhile, the British tried to retaliate. On December 15, the same day that Bush had sent the new policy to the Military Policy Committee, Conant asked the British to send Professors Chadwick and Peierls to the United States to confer with Robert Oppenheimer.[33] It was clear to Akers that the Americans were interested in extracting information on the diffusion process from the two Englishmen.[34] On January 11, Anderson passed on the same sentiment to Churchill.[35] Throughout January Groves and Conant repeated their request, but on every occasion they included the rider that the exchanges did not imply that there would be a continuation of exchanges. Chadwick and Peierls would have to convince Groves that further exchanges would be of assistance to the Kellogg Company, the American firm working on the diffusion plant.[36] Chadwick and the others were insulted. They turned down the invitation and recommended that no helpful information or progress reports be sent to the United States.[37] Michael Perrin, Akers's deputy in England, went further by suggesting that they should retaliate by working with the Canadians and building their own plant in England.[38]

However, before any drastic steps could be taken, Churchill sought a diplomatic solution. The Roosevelt-Churchill Casablanca Summit opened on January 12, 1943. The British scientists were optimistic that both leaders would resolve the atomic cooperation problem.[39] They did not. Churchill was reluctant to raise the issue directly with the president. Both leaders had too many more immediate war-related matters to discuss. Their attention was held by, among other matters, Allied operations in North Africa, preparations for the invasion of Sicily, German setbacks outside Stalingrad, and the persistent demand of Stalin, not in attendance, for a second front in Europe. Churchill, instead, appealed to Harry Hopkins, Roosevelt's trusted adviser and confidant, on January 18. Originally from Iowa, Hopkins had made a reputation for himself as a social worker and social administrator in New York. He had worked very closely with the president in implementing the New Deal. Although Hopkins held no official cabinet post, the president trusted him and often used him for special diplomatic missions. He was widely disliked both by the president's political opponents and by supporters who questioned his influence over Roosevelt. In fact, he actually had his own bedroom at the White House, and when he announced his plans to get married, Roosevelt invited him to move his wife in. Churchill, recognizing the president's close ties with Hopkins, had already established a good relationship with him. Hopkins assured Churchill that the president knew exactly how to handle the British concerns and that "it would be entirely in accordance with our [British] wishes."[40] However, Hopkins, if he knew, did not tell Churchill that the president had approved the new policy.

On January 25, after returning from the conference, Churchill sent Hopkins a reminder, and on February 15 he asked for some news, "as at present the American War Department is asking us to keep them informed of our experiments while refusing altogether any information about theirs."[41] On February 24, Hopkins requested a full memorandum on the "basis of the present misunderstanding" because "our people feel that there has been no breach of agreement." Three days later he had his memo, which, among other things, again mentioned the conversation between both leaders at Hyde Park in June 1942, where it was agreed, Churchill wrote, "that everything was on the basis of fully sharing the results as equal partners. I have no record, but I shall be surprised if the President's recollection does not square with this."[42] This is the same conversation alluded to by Churchill in *The Hinge of Fate.*

On the very day of the prime minster's response, Anderson received a request from Conant to have Halban hold discussions with Urey and Fermi in the United States. The Americans wanted to discuss his original work at Cambridge on heavy water. Mackenzie viewed the request with optimism, and so did Halban; both men interpreted it as a sign that the Americans were reconsidering their stand on collaboration.[43] Anderson, however, had other ideas. He ordered that future contacts with the Americans should be limited until Churchill and Hopkins worked out future arrangements for cooperation.[44] Two days later he instructed Halban "to find a suitable pretext for postponing these discussions."[45] The British welcomed this opportunity to send a strong message to the Americans. Halban, very reluctantly and after futile appeals to London, agreed to contract a "diplomatic flu." But by then, London was not satisfied with a claim to illness. It instructed him to state quite frankly that he was acting under instructions that were due to the United Kingdom's dissatisfaction with U.S. policy. Urey was infuriated with Conant for the cancellation of the meeting. He accused him of putting nationalistic concerns ahead of the completion of the bomb. Lack of cooperation, he argued, could delay his heavy water investigation by six months to a year.[46]

Meanwhile, Churchill sent another reminder to Hopkins on March 19. The letter jolted him into action; apparently, he had been sitting on his hands. Hopkins gave a copy of the February 27 memo to Roosevelt, who in turn sent it to Bush on March 24 with instructions to prepare a reply. Bush and Conant huddled and came up with a reply to Churchill's "assertion of a breach of agreement," which was dispatched to Hopkins on March 31. The reply consisted of two memos, one written by Conant and the other by Bush. These documents are revealing, for they go a long way in explaining why Bush and Conant steered the president to adopt limited collaboration. Bush's memo repeated the argument that the British did not propose to build atomic energy plants in England during the war. Hence, giving them information that they did not plan to use increased the risk of security leaks.[47]

However, the most revealing of the two memos was Conant's. He discussed at length two issues that were clearly of paramount importance to him: (1) the postwar political and military potential of the weapon and (2) the postwar industrial application issue. As Conant saw it, the British government, by working so closely with Imperial Chemical Industries (ICI), had given that company an advantage over all others, including those in the United States, for postwar industrial exploitation of atomic energy. Whereas in the United States no one with industrial connections was in a policy-making position, Conant observed that two of the most important British policy makers—Akers and Perrin—were on leave from ICI to work specifically on atomic energy. Indeed, he opined, the entire controversy over collaboration might never have arisen if negotiations had been left to British scientists "and if those British scientists had had the same voice in determining policy in Great Britain as you [Bush] have had here in the United States."[48] He had no doubts that the British were aware of the industrial potential of atomic energy. He quoted the MAUD Report, which had stated that the new sources of power would "affect the distribution of industry over the world"; hence, Britain must "take an active part in this research work so that the British Empire cannot be excluded by default from future developments."[49] Having stated his case, Conant added that the possible benefit to American industry was, in his view, a very minor consideration. "The major consideration must be that of national security and postwar strategic significance."[50] In terms of the security of the United States, he explained, "knowledge of the design, construction and operation of these plants is a military secret which is in a totally different class from anything the world has ever seen. . . . Therefore, the passing of this knowledge to an ally under conditions whereby the ally cannot profit directly in this war would seem to raise a question of national policy comparable at least to alienation of control of a fortress or strategic harbor."[51] He thought that it was his and Bush's duty to make the president aware of what was involved before he made any new decisions.

At least one Englishman had already anticipated Conant's concerns about ICI. Cockcroft did foresee problems with the United States over the ICI scientists who worked on the British project. But, as he pointed out, almost all the available scientists and engineers capable of working for the government had to come from private industry. He noted, however, that ICI people were in fact too highly placed and that they continued to maintain close contact with the company. This association, he observed correctly, would lead the Americans to "suspect that the British were trying to steal a march on them in the development of industrial applications after the war."[52]

Unaware that Hopkins finally had taken some action, Churchill prodded him again on April 1.[53] There was very little Hopkins could do; Bush and Conant's memos had left a strong impression against improved collaboration. Hopkins could only promise to send a telegram explaining the matter. No telegram arrived. The British had grown tired of waiting. In April Perrin suggested that they and the Canadians should present a united front to the Americans and demand collaboration. If they failed to get a favorable response, then both countries should consider a joint program.[54] At the same time, Cherwell proposed a feasibility study on an independent British program. Churchill too, on April 15 requested "a rough estimate of the financial cost and the manpower required. . . . In my view, we cannot afford to wait."[55] The prime minister indeed did not have long to wait. Akers and Perrin had already done a study that projected that the cost would be astronomical. Nonetheless, the Technical Committee endorsed it and even recommended an expanded program that would include the construction of a full-scale plutonium plant in Canada.

If Churchill had agreed to proceed with an independent British program or even an Anglo-Canadian endeavor, that decision certainly would have changed the future course of atomic research, though to what extent is anyone's guess. Perrin later admitted that an independent project would have been no more than research work. In addition, he recognized that if the United Kingdom had gone into competition with the United States, the Canadians, however much they might have desired to help, would not have been in a position to give any effective help due to their special relationship with the United States.[56] In fact Dean Mackenzie had already warned Akers in mid-May that if the Canadians were forced to choose sides, "they will undoubtedly refuse to take any action which will antagonize the American Government."[57]

The matter was still under consideration when Churchill left to attend the Trident Conference in Washington with Roosevelt. On May 15, three days into the conference, Anderson sent him some disturbing news. He had learned that the U.S. government had signed a contract to purchase the entire output of the Canadian uranium mines for the next two years. Additionally, the United States had bought up the entire production of Canadian heavy water. Anderson's note continued: "It seems, therefore, that if we break with the United States on tube alloys, we shall be deprived of our only source of uranium and any early possibility of delivery of heavy water."[58] He recommended strongly that the prime minister make every possible effort to bring about effective cooperation. This revelation about American monopolistic activities in Canada made moot any further debate on a British independent program. It was now completely unthinkable, as it was not feasible to pursue technical experiments without a reliable source of raw material. This development marked the first of many instances when the availability of atomic raw material influenced the policies of the British and Americans.

Churchill was quick to grasp the gravity of the British situation and the disadvantageous position that the British were in. He raised the question of cooperation directly with Roosevelt. The president readily arranged a conference for May 25 to enable Bush to justify U.S. policy to Lord Cherwell and Hopkins. Bush was none too happy, as he would later inform the British, that the problem was raised at the highest level. At the meeting Cherwell denied American charges that the United Kingdom had only a postwar commercial interest in atomic energy and none in the development of a bomb for wartime use. However, he did admit that the British wanted to stay abreast of the technology so that after the war they would be in a position to develop their own weapon. Hopkins was a keen observer; he confessed that for the first time he understood what the issues were. Bush's impression was that Hopkins "felt that the British were asking for something on which they had not the slightest basis for the request."[59] Hopkins reported to the president that collaboration should be resumed. The following day Roosevelt and Churchill agreed to resume collaboration. Churchill, confident that the problem had been resolved, cabled Anderson to that effect.[60] His optimism was somewhat premature.

The initiative again ran into a roadblock. Seven days later, the president had a luncheon with Bush at which, one may speculate, he planned to resolve the matter. Bush reported that at the meeting he had made the disclosure to the president that Lord Cherwell had intimated that the United Kingdom had placed the whole affair of atomic energy on an after-the-war military basis. The president, in Bush's words, "agreed that this was astounding." Bush stated that several times during the course of the conversation the president came back to the U.K. position and that on every occasion he expressed his amazement at it.[61] The president added, in response to Bush's inquiry, that he had not received any "report" from Hopkins on the conference with Cherwell. This implies that Hopkins had kept him in the dark about what had been discussed. Roosevelt has been known to act uninformed in order to gauge other people's thoughts. One cannot dismiss the possibility that Hopkins might have discussed with him the conference but not the issue of the British's postwar position. One may add that Hopkins had indeed asked Bush to keep the nature of the discussion quiet.[62] Assuming that the information was new to Roosevelt, obviously he felt that it was inappropriate to introduce his new policy to Bush at that time. Bush left the luncheon convinced "that the President has no intention of proceeding farther on the matter of the relations with the British."[63]

The British, of course, knew nothing about all this, but there was a suspicion that something was amiss. Six weeks after the conference there was still no movement. In the interim, Churchill had sent off a reminder to Hopkins. It was uncharacteristic of the prime minister to beg, hat in hand, for anything. Throughout his political career he had cultivated a well-earned reputation for arrogance and pride. But faced with the demands of his atomic advisers and the precariousness of the British position, he had to swallow his aristocratic pride. On July 9, he rushed off another telegram, this one to the president: "Since Harry's telegram of 17th June I have been anxiously awaiting further news about Tube Alloys. My experts are standing by and I find it increasingly difficult to explain delay. If difficulties have arisen, I beg you to let me know at once what they are in case we may be able to help in solving them."[64] Roosevelt was now faced with a serious predicament. Should he ignore this telegram as he had ignored the others and hope that Churchill would drop the issue? Or should he disdainfully ignore the advice of those who had counseled against collaboration and resume interchange? The former option was out of the question. He and Churchill were scheduled to meet in Quebec in five weeks to discuss the progression of their strategic programs against the Germans. It was inevitable that he would have to fend off Churchill's charges of breaking his word on the atomic question. Since the issue meant so much to the British, there was a likely chance that if left unresolved it could affect the resolution of other war-related matters. One of those matters involved the invasion of Hitler's Europe. Roosevelt and the Joint Chiefs were resolved to convince

Churchill to give priority to a cross-channel invasion as opposed to his preference to knock Italy out of the war and advance on Germany from the Mediterranean. Moreover, the Americans wanted the operation to be placed under an American commander. Roosevelt had to be conscious of the potential effect of the atomic energy problem on American efforts to sell their proposals to a disgruntled Churchill.

On July 14, Roosevelt asked Hopkins for advice on the atomic issue. Hopkins's reply was unequivocal: "I think you made a firm commitment to Churchill in regard to this when he was here and there is nothing to do but go through with it.[65] On the same day, July 20, that he received Hopkins's memo, Roosevelt wrote to Bush:

Dear Van,

While the Prime Minister was here we discussed the whole question of exchange of information regarding tube alloys, including the building project.

While I am mindful of the vital necessity for security in regard to this, I feel that our understanding with the British encompasses the complete exchange of all information.

I wish, therefore, that you would renew, in an inclusive manner, the full exchange of information with the British Government regarding tube alloys.[66]

The president had finally stepped off the fence to make a decision, one that was quite painful to him since it forced him to adopt a policy that was contrary to the best advice of his experts in the field. Unfortunately, from the British standpoint, Bush would not learn of this letter for ten days. He and Stimson had already left for London to hold discussions on various issues with the British.

Anderson had no intention of seeking out Bush for a meeting. He was not optimistic that Bush's visit would bring about any changes.[67] On the other hand, Perrin, out of desperation, thought that nothing could be lost by initiating some sort of contact. On his own initiative he managed to "scrounge" a personal invitation to a Royal Society reception that he knew Bush would be attending. Perrin's meeting with Bush at this reception would set in motion the chain of events that would lead to the Quebec Agreement. At the reception he managed to corner Bush for a few minutes of private conversation. Bush informed him, off the record, that the present difficulty had arisen principally because the postwar implications of atomic energy had become a factor and that the president could not make postwar commitments without the approval of Congress. He pointed out that as the president and Congress were at war with each other, this made the position even more difficult. Bush then revealed for the first time that the British had made "the greatest tactical error" by going behind his back and that of the army and the secretary of war to appeal directly to the president. Everyone was offended, he explained, and the consensus was that everything should be done to resist the United Kingdom's proposals to the president.[68] Perrin informed Anderson of the conversation.

Churchill then arranged a session with Bush, where, in Bush's words, "the matter was raised powerfully, to say the least."[69] Churchill stated that the president had given him his word of honor that both nations would share equally in the efforts but that "every time an attempt was made to modify the present arrangements it was knocked out by somebody in the American organization."[70] In response to the postwar commercial charge, Churchill responded that he "did not care about any post-war matter" but that he wanted to be in shape to handle the affair in the present war and that alone.[71] Bush had hoped that collaboration would not become an issue while he was in England. However, if the issue were raised, he and Conant had already agreed to pass the buck to Stimson. Bush told Churchill that the matter was not in his hands but Stimson's. He and Harvey H. Bundy, Stimson's special assistant, briefed the secretary of war, who subsequently had a private conversation with Churchill on July 17. Stimson explained that it would be difficult for the United States to share the commercial fruits of atomic energy after the war, especially when Congress found out that American dollars, 500 million to date, had been spent on the project without its knowledge. He confessed that the president had told him nothing about the discussions held by both leaders on atomic energy and that he was disturbed to know that Churchill thought the British had been unfairly treated. He told Churchill that there was someone high up on the United Kingdom's side who had been explaining forcibly that the United Kingdom's interest in the matter was commercial and after the war.[72] To the chagrin of Bush, Stimson left the meeting with "a great deal of sympathy with the British point of view." Bush tried to persuade him otherwise, but Stimson "felt very strongly that the present situation must be altered in the direction of full interchange."[73] To no avail, Bush and Bundy went over the many obstacles to a postwar agreement in an effort to change his view. By a remarkable coincidence, Stimson's opinion in favor of renewing collaboration was formed three days before Roosevelt had decided to send Bush his letter to that effect.

By agreement, Stimson and Churchill decided to hold a formal meeting on July 22. But even before the participants came together, Churchill was already entertaining the idea, which he shared with Anderson, of conceding all commercial rights to atomic energy to the United States.[74] He received no arguments from Anderson, who believed that the necessary rigid control of atomic energy would prevent its application to industrial use.[75] Cherwell too supported his prime minister's suggestion because he did not believe atomic energy would ever have economic uses.[76] Consequently, Churchill's idea was incorporated into the draft minutes to be pursued at the meeting. Perrin protested, but he could not move Anderson and Cherwell to change their strongly held view that atomic energy had no future industrial application. He wrote Akers, who was in Canada: "I think you will share our [the Technical Committee's] feelings over here, that this is not in fact the case, and I cannot help feeling that there are many Americans who would agree with this."[77]

The meeting was held at 10 Downing Street. In attendance were Churchill, Anderson, and Cherwell on the U.K. side and Stimson, Bush, and Bundy on the U.S. side. Churchill summarized the British position by touching on his agreements with Roosevelt, the Conant Memorandum, its effect on future Anglo-American relations, the risk that Germany or the Soviet Union would perfect a bomb, the unrealistic U.S. belief that it alone should control the bomb, and the United Kingdom's determination to start a parallel bomb project. He repeated that the United Kingdom had no interest in the commercial aspects of atomic energy. In his response Bush defended "need to know" as a necessary security device that served the best interest of both sides. At his mention of the postwar commercial aspect, Anderson interjected that the commercial issue had confused matters and that probably the United States "had used the commercial possibilities as a camouflage for the real purposes in the effort."[78] He did not mention the "real purpose," but undoubtedly the reference was to military power in a postwar world. This was an astute observation on the part of Anderson that seems justified when one considers the many reasons Bush and Conant had proposed, at different times, to justify noncollaboration. Of all the reasons advanced, the commercial aspect had emerged as the most credible; hence, Bush and Conant had ridden it for all it was worth. The ride was about to end.

Churchill proposed to the Americans a five-point agenda between himself and Roosevelt that should be used as the basis of an agreement. The five points were

1. Completely joint enterprises with free exchange of information
2. That each government should agree not to use the bomb against each other
3. That each government should agree not to share information with third parties without the consent of both
4. That they should agree not to use the bomb against any other parties without the consent of both

5. That the commercial or industrial uses of Great Britain should be limited in such manner as the president might consider fair and equitable in view of large additional expense incurred by the United States[79]

By agreeing to article 5, the British had voluntarily sacrificed the postwar commercial rights to atomic energy in order to gain American agreement to a collaboration effort that, unbeknownst to them, the president had already decided to grant. Their acquiescence also served to take away Bush's commercial rationale for noncollaboration. The following day Bush rushed off a memo to Conant, briefing him on the new developments. He added: "I have wished many times in the last week that you and I could sit down and puzzle things out. I have had to go along for there was no way whatever of stopping the progress nor would it indeed have been advisable to do so, and I merely hope that I have not made a lot of mistakes in advising the Secretary."[80] Around the 27th, he received a cable that conveyed the president's July 20 instructions to review full interchange. Upon receipt of the cable, he had assumed that Stimson had communicated the results of the July 22 meeting to the president and that the president was instructing him "to review with the British the matter of interchange" as a follow-up to the just-concluded meeting. This confusion was not cleared up until after he returned to the United States and saw the original letter.[81]

Meanwhile, Conant was beside himself with rage, part of it directed at Bush, when he learned of the agreement. In his response to Bush, written after he learnt of the president's reversal, he wrote: "May I record the conviction expressed in previous memoranda in your files that a complete interchange with the British on the S-1 project is a mistake. . . . I can only express the hope that the President did not reverse his decision on a matter which may have such important bearings on the future of the United States without proper understanding of the potential possibilities of the weapon we are now engaged in developing, nor the difficulties of our enterprise."[82] Revealingly, the memo, written in the first throes of disappointment, made no mention of the commercial factor. Four days later, in a handwritten note in which he made the observation that the president and the team in England had arrived at the same conclusion independently, Conant expressed sentiments that seemed to imply that Bush had not done enough to derail the agreement. "It's not for the likes of me to question your judgment but I'd like my memo to stand for the record none the less. Perhaps I'd better join the staff of the Chicago Tribune [where his Anglophobia would be more at home?]."[83] By August 6, he had simmered down considerably. He had studied the proposal and now felt "that a renewal of the interchange under these conditions is tantamount to an acceptance by the British of our original offer. . . . It seems to me that we can now proceed with an interchange which will be in the best interest of the United States and the war effort."[84] Moreover, he pointed out, because the United States was at a point where it needed all the help it could get on the diffusion process, "the renewal of interchange comes at a moment which will make the interchange of special value to the American effort."[85] Conant had turned out to be his own best spin doctor, although, as developments would show, his assessment of the proposal was absolutely correct. In their haste to grasp at the straw of collaboration, the British had tendered a proposal that, on interpretation and execution, turned out to be a fragile framework for meaningful collaboration.

One of the most astonishing things to come out of the July 22 meeting, to paraphrase Perrin's observation, was Bush's claim that he had never seen the Conant Memorandum and his admission that he thought it to be "a most unreasonable document." It was another attempt by Bush to divorce himself from the process and the decision makers responsible for drawing up the restrictive memorandum. However, on this occasion the British were not fooled. The files of the Tube Alloys Directorate were searched immediately, and Akers was contacted by telephone in Canada to help clarify the matter. The prime minister was informed of the results on the 23rd. The findings disclosed that Conant gave the memorandum to Akers on January 13 in Bush's office. Akers corroborated that Bush was present when Conant read the memorandum aloud to Akers. Akers added that even if Bush was not present on January 13, he had on other occasions discussed with Bush the contents of the memorandum.[86] Anderson knew that Bush had lied but advised Churchill that they should let him down lightly and concentrate on cultivating his support in Washington. He believed that Bush and Stimson would be needed to counter General Groves's opposition.[87] As it turned out, Groves supported the agreement but for his own reasons, as we shall soon discuss.

On August 19, while attending the Quebec Conference in Canada, Churchill and Roosevelt signed what became known as the Quebec Agreement.[88] To ensure "full and effective collaboration," a Combined Policy Committee (CPC) was created. Its membership was composed of three U.S. representatives, two British and, "in view of the part which the Canadian Government are playing in this project,"[89] one Canadian representative. The Quebec Agreement was a secret arrangement, the terms of which would be known to only a few insiders on both sides of the Atlantic. For years after the conclusion of the war, its very existence would be unknown to the British Parliament and the American Congress. And even after its existence came to light, steps would be taken by both governments to keep its terms secret.

Did wartime considerations really play a role in the outcome of the atomic debate? Hershberg argues that when Roosevelt gave to Churchill, at the Trident Summit in May 1943, his assurances to resume full atomic collaboration, he did so because "he desired a harmonious Anglo-American alliance and a smooth relationship with Churchill." The war was far from over, and both leaders still had several momentous decisions to make.[90] In other words, Roosevelt made atomic concessions to Churchill in order to cultivate an amicable atmosphere for the conduct of the war. Along the same lines, Sherwin's view is that Roosevelt's concession "marks the point at which the President began to deal with atomic energy as an integral part of his general diplomacy, linking and encompassing both the current wartime situation and the shape of postwar affairs."[91] That is, the reasons for the concessions went beyond the realm of the immediate war to the postwar period. He and Churchill were constructing an atomic alliance that would continue after the conclusion of the present war.

It is a bit premature to identify the germination of this linkage as early as May 1943. True, both men already appreciated the potential power and political implications of the new weapon. But the circumstances surrounding Roosevelt's concession do not support the argument that he was thinking of a postwar atomic alliance. Roosevelt had clearly accepted the arguments of his atomic advisers that the Americans could and should proceed with an independent project. But Churchill would not be ignored. After the concessions at Trident, the president still resisted, forcing Hopkins to remind him that he had given his word to Churchill. If the Quebec Agreement itself hinted at an atomic partnership beyond the war, one should mind that Roosevelt had no input in drawing up the articles of the agreement. The Quebec Agreement was primarily a British agreement. So if anyone was thinking of a postwar atomic alliance, it was Churchill, not the president. Churchill had not yet won over, if he ever did, the president to his views, which were in their infancy, of using the weapon as an instrument in postwar diplomacy. More likely, the questions surrounding the present war and the imminent prospect of raising them with a disgruntled Churchill swayed Roosevelt to renew collaboration. At Quebec, Churchill gave his conditional assent for a 29-division cross-channel operation with a target date of May 1, 1944. He agreed even though his sentiments were still in favor of a Balkan invasion, where he thought capitulation would follow as quickly as in Italy, where Benito Mussolini had fallen on July 25. It is quite likely that the president's concession prevented the atomic energy question from poisoning the waters of military cooperation. Given how exasperated Churchill had become over the atomic collaboration issue, it is not far-fetched to speculate that if he had not gotten his way, Churchill might have hardened his resistance to the invasion of France.

The British finally had won the collaboration they had so desperately sought. But, as time would reveal, it was a hollow victory that offered less rewards than they thought. Groves, Bush, and Conant had taken great pains to construct an "American" atomic project, the benefits of which were to be used to promote American interest. They were convinced that they could construct a bomb without British help and had conveyed their conviction to the president. Despite their assurances, the president, through a combination of military considerations, friendship, and quite possibly honor, agreed to atomic cooperation. But although British scientists and politicians were enthusiastic and optimistic about the future opened up by the agreement, the president's advisers were still not willing to share fully their expertise and technical know-how. Simply put, the British were invited to assist the Americans in the development of an American bomb, and if they hoped to continue contributing to the project, they were expected to toe the line and observe American rules. The British were prepared to accept the restrictive conditions imposed by the Americans because, though willing, they were unable to proceed independently.

THREE

U.S. Bureaucrats Limit
Post-Quebec Collaboration

The Quebec Agreement failed to usher in a smooth period of Anglo-American collaboration. Throughout the remainder of the war, areas of disagreement arose, and always the Americans imposed their will upon the British. Roosevelt's atomic energy advisers had agreed to cooperate only because the policy was thrust upon them from above and by the unrelenting pressure and determination of the British government. However, the one thing these advisers were not willing to do was accept the British as equal partners. Beyond any doubt, the British accepted this secondary role. They justified this acceptance by confessing that they harbored a fear of offending the Americans, thereby giving their "anti-British American enemies" an excuse to pull the rug of cooperation from under their feet. Almost from the moment of the signing, they began bending to the demands of the Americans. Their yielding strategy cost them the respect of the Americans, who tended to view the British as tiresome partners to be tolerated for the duration of the war or the completion of the bomb, whichever came first. The problem was not destined to end that soon. Roosevelt and Churchill again held one of their private one-on-one summits, and what came out of that meeting was destined to have far-reaching effects in the postwar era. Meanwhile, Groves introduced an issue that played just as important a role in influencing postwar Anglo-American relations: the quest for atomic raw materials. Uranium surfaced time and again over the following years to play a dramatic role in the roller-coaster atomic energy relations between the United States and the United Kingdom. Hence, it is worth looking at the early history of uranium acquisition and Anglo-American relations in this area.

As previously mentioned, Groves, like Conant, quickly recognized how the Quebec Agreement could be used to American advantage. The Quebec Agreement fitted right into his plans to secure a worldwide monopoly of raw

55

materials so that, as he put it, "in the future the United States would not lack the essential raw materials . . . suitable for the production of atomic energy."[1] This is not to say that he was happy over how the agreement was negotiated. As late as 1950 he was still complaining that neither he nor anyone else associated with the Manhattan Project knew anything about the agreement until after the details were worked out.[2] Even before the Quebec Agreement was signed, as noted earlier, Groves had taken steps to corner the Canadian raw material output. The United Kingdom, under Anderson, had tried to negotiate a discreet takeover of Eldorado, the Canadian uranium company, as early as May 1942. Anderson had proposed that the Canadian government should acquire financial control of the company, to avoid competitive buying, and that the output of the mine should be divided between the British and Americans.[3] He had shared this idea with Bush.[4] Hence, when he discovered that Groves had practically monopolized the Canadian output of uranium and heavy water, he felt that the United States had stabbed the British in the back.[5] As it turned out, by ordering 850 tons of ore over the next three years and arranging for Eldorado to refine a consignment of Belgian ore that Union Minière had transferred to the United States on the outbreak of war, the United States had tied up the mine and refinery for its purposes for five years.[6]

Groves did not stop with this successful maneuver. In the fall of 1943 he sent one of his geologists to the Congo to locate new sources of uranium. What the geologist found was that the mines at Shinkolobwe contained some low-grade uranium ore—3 to 20 percent content. Only the richest ore—60 percent content—had been sorted by hand and exported from the mine.[7] Groves approached Edgar Sengier of Union Minière, who refused to assume responsibility, in the absence of the controllers of the company, for committing all the Congo ore to the United States or giving the United States the right of first refusal on all ores produced. However, negotiations ensued on the purchase of a specific tonnage of newly mined ore. Price became the stumbling block.[8]

Meanwhile, faced with the loss of the Eldorado supply, Colonel J. J. Llewellin, a U.K. member of the Combined Policy Committee (CPC), newly created by the Quebec Agreement, suggested that the United Kingdom attempt to corral the Congo ores for British use. Anderson feared that the Belgians might take advantage of any such attempt to engineer a bidding war—something that Sengier had no plans to initiate—so Anderson vetoed the suggestion. He knew that Britain could not win a bidding war with the United States. Besides, although the Quebec Agreement had made only vague references to the allocation of "materials . . . in limited supply," Anderson preferred a collaborative effort that would not jeopardize the agreement.

Groves was not aware of Anderson's preference, but he was not willing to risk losing the Belgian ores. In the face of Sengier's intransigence, Groves decided to turn to the CPC. The major advantages of this move were Britain's long-standing close relationship with the Belgian government (the colonial masters of the Congo) and the investment of British shareholders, who owned 30 percent of Union Minière. Groves later explained his decision in the following terms:

> The Military Policy Committee felt that if Great Britain took advantage of the location in London of the Belgian government in exile or, later, of the normal British strong influence over Belgian policy, it could and would secure a monopoly over the Belgian Congo raw materials. The United States would then be in a most disadvantageous position. . . . It was realized that our best prospect for obtaining an exclusive long-term commitment from the Belgians would be through the medium of a government agreement between Belgium on the one hand and the United States and the United Kingdom on the other.[9]

The CPC then agreed that a tripartite agreement should be negotiated between the United States, the United Kingdom, and the Belgian government in exile in London. The end result of the negotiations, which were conducted by Anderson and John Gilbert Winant, the American ambassador to the Court of St. James, was the Agreement and Declaration of Trust that was signed on June 13, 1944, by Roosevelt and Churchill. The trust agreement established in Washington an agency known as the Combined Development Trust (CDT), which was placed under the direction and guidance of the CPC. In effect the CDT became the raw material acquisition and allocation arm of the CPC. The main function of the CDT was "to secure control and insure development of uranium and thorium supplies located outside the jurisdiction of the United States, the United Kingdom, the Dominions, India, and Burma."[10] In other words, its main responsibility was to establish a worldwide monopoly of all known sources of raw material.[11] The long-term result placed the British and the Americans as equal partners in the largest known source of uranium in the Congo. This turned out to be very significant in the postwar period.

Other issues quickly arose to illustrate the tenuousness and inequity of the Anglo-American atomic partnership. The ink had hardly dried on the Quebec Agreement when the English invasion began—to the consternation of Bush and Groves. Akers arrived from Canada before it was signed, and he cabled London to send over Chadwick, Oliphant, Simon, and Peierls. They arrived on the very day the agreement was signed. When Bush found out that Akers had sent for the four men, he told him "quite forcibly" that the action was premature because the CPC would first have to meet to set up rules and procedures and determine where and how to use the British scientists. He warned Akers that the four men would be left sitting around doing nothing; hence, he hoped that Akers would make it clear to them that the United States had nothing to do with their coming.[12] Whether Akers made that clear to the four scientists is uncertain, but they certainly sat around doing nothing for a considerable period of time. He made the rounds but to no avail. Conant told him that, personally, he supported the restoration of collaboration but that "it was essential that we should not try to force the pace."[13] He was advised to wait until the CPC met on September 8.

Conant also told Chadwick that at that same CPC meeting the U.S. representatives would produce a report, which had recently been prepared for the president, showing the extent of the American effort and their time schedule for large-scale work. He advised Chadwick that he should criticize this report as much as he could, since he felt that this would be the best way of showing the American members of the CPC that the British had much to offer and that it would make them inclined to more extensive collaboration. It is not clear whether Chadwick took this advice, which obviously, as Conant was well aware, would have caused acrimonious feelings at the very first CPC meeting. But it is clear that the atomic advisers were alert to finding and exploiting any obstacles to collaboration.

Churchill and Akers had hoped that the president's advisers would not resist. The prime minister was not pleased with the Quebec Agreement. But he was absolutely sure that the British could not have received better terms and that only those who knew "the circumstances and moods prevailing beneath the Presidential level" would understand. He added, "I have no fear that they will maltreat us or cheat us."[14] Akers was not too happy with the agreement either; he realized that its success depended almost entirely on American "good-will and active participation at the beginning." In addition, he knew that the only way the British would be able to build a plant after the war was if their engineers worked with the Americans on the design, erection, and operation of their plants. Nonetheless, he was optimistic that once collaboration got properly started things would run of their own accord, "unless there is active, or even passive, resistance from Bush, Conant and Groves."[15] There would be such resistance.

On September 10, Groves called a meeting of the S-1 Executive Committee, at which he informed them of the signing of the Quebec Agreement. He then took the first step in ensuring that the British contribution to the Manhattan Project would be kept a secret. He stated that no mention of Anglo-American relations should be made outside of the members of the S-1 Executive Committee unless on official orders. Furthermore, he warned them that no one should exchange information with a British representative except acting under his specific orders.[16] Thus, while Akers and his team were anxiously waiting out the delay, Groves had already begun taking steps to limit free interchange. He did not have to look far beyond the agreement itself to find the means of limitation. The United States had always sought, before and after the Quebec Agreement, to make a distinction between basic, scientific information and technical, manufacturing "know-how." It had been more willing to furnish the former than the latter. This distinction was clear in the Quebec Agreement, and it is surprising that the British accepted it. The agreement provided for "full and effective interchange of information and ideas" in the field of scientific research and development—and only between those engaged in the same section of the field. In the area of "design, construction and operation of large-scale plants, interchange of information and ideas shall be regulated by such *ad hoc* [underlined in agreement] arrangements as may . . . appear to be necessary or desirable."[17] When both Conant and Groves decided to support the agreement, it was because they both recognized that it provided the British with not much more than the Conant Memorandum and that they would be well positioned to control the process leading to the "ad hoc" arrangements that the accord envisioned.

Resistance emerged in other areas as well. No one on the American side trusted Akers, simply because he, as well as Perrin, was on loan to the British from Imperial Chemical Industries (ICI), the British chemical plant that alone among English companies had the experience to work with Tube Alloys. This concern had been raised before, but this time it was taken to the highest level. While still in Canada, Roosevelt showed Churchill a memo from Bush asking that Akers be replaced as Anderson's chief atomic liaison. Bush explained that in the past there had been difficulties because "the British representative was an industrialist." Hopkins then asked Churchill most earnestly that "Bush's reference to Akers should not be [*sic*] any conceivable chance reach eyes of the latter."[18] Churchill did not share Bush's view, nor did Anderson. Nevertheless, in the interest of collaboration, they decided to replace him as technical adviser with Chadwick. Akers agreed to remain in charge of the overall Tube Alloys project, but by November it was apparent that he had lost his enthusiasm for the whole project.[19] All the same, with the resolution of the Akers matter and the meetings of the CPC and its subcommittees, collaboration started off under American terms—severely restricted by Groves's compartmentalization policy.

Why did the United Kingdom agree to compartmentalization? British representatives were aware of the policy long before the signing of the agreement and had protested as far back as the summer of 1942. Yet when Anderson went to Washington to work out the details of the agreement, he accepted a draft that gave the policy an official stamp of approval. We know that Akers had convinced himself that Groves's tight control would be relaxed as collaboration progressed. Did both men also believe that the policy would affect primarily American scientists and not members of the British team? As early as October 1943, Akers had written to the Canadian-based team to share his observation that the system had been carried "to a point at which efficiency is seriously hampered." But, he advised, it was their system, and "it is essential that we should not be responsible for breaking down their system." It was the opinion of the British team, he explained, that the American scientists were hurting themselves by not sharing information with each other. He advised the team that it was extremely important that members not disclose to the American teams any information that was withheld from the Americans. He admitted that it was not going to be easy because the Americans tended to "pump" the British workers in order to learn more about the project in general.[20] Does all this mean that Akers believed that the British would occupy a special position—that they would be given the freedom, which Americans did not have, to move from one compartment to another? It is possible that he did believe so.

Dr. Mackenzie, head of the Canadian Research Council, revealed in his diary another explanation for the British's acceptance of the system: "It is apparent that Chadwick feels as I have always felt, that the United States effort is one hundred times greater than any possible United Kingdom effort, that the Americans can get along if necessary without the UK, while the UK can do nothing without the US; and that our best policy is to make sure we get collaboration and also that we will have to change our plans to suit the American situation."[21] The British had grown desperate during the dry months of noncollaboration. Their leader had begged, threatened, and cajoled the Americans for a resumption of interchange. Anderson knew of every letter and memorandum sent by Churchill; he knew what it had taken to get the talks going again. He had no intention of derailing an agreement by raising contentious questions over "need to know." Akers's assurances to him that the restrictions might be temporary only served to weaken his resistance even more. Moreover, not every Englishman was critical of Groves's security measures. H. C. Webster, a British scientist, confessed to Perrin that he had "a great deal of sympathy with Groves's attitude on security." He recognized that gossip and publicity could be disastrous to the project. It would generate questions in Congress, and some in the United Kingdom might question the morality of the project.[22] In January 1945, by which time it was clear that the system also applied to British scientists, Chadwick confessed that he understood the rationale for Groves's restrictions on U.K. personnel visiting U.S. industrial laboratories and other research institutions. Groves, said Chadwick, feared that the visits could generate awkward questions and enquiries that could lead to loss of security. He continued, "I am giving this explanation because I feel Groves's arguments may seem somewhat farfetched, but there is some substance in them."[23]

As it turned out, compartmentalization restricted the free flow of information among all groups—American and British alike—working on the Manhattan Project. The British were almost totally cut off from areas dealing with the large-scale construction and operation of plants. In addition, strenuous efforts were made to isolate them from all plutonium work. In a 1989 interview, R. Gordon Arneson, the recording secretary of Truman's Interim Committee on atomic energy, still boasted that "we'd kept the British away from plutonium altogether. Among their 24 scientists who were sent over here, none ever got to Hanford at all, and that was deliberate. We didn't want them to know much about plutonium."[24] The result, of course, was that after the war the British scientists returned to England with a piecemeal knowledge of the overall project. Their exposure to narrow areas of the project and its potential effect on their postwar plans were not lost on the British. In early 1945, Chadwick commented to Anderson: "Our participation in this project represents a very small fraction indeed of the total effort and it will not be immediately obvious to Congress and the American public that we have taken any significant part. We shall have to rely very much on the US authorities, and especially on Groves, to work with us towards collaboration, not merely to accept collaboration if it is forced on them."[25] Chadwick could not have been more perceptive. Groves would hold the key to immediate postwar collaboration.

Unfortunately for the British, Groves had already decided against postwar collaboration. But they did not know this. Groves astutely led the British to believe that he supported postwar collaboration. Early in the year, Groves had received word that the British had started work on certain atomic energy plants, and he wanted to be kept fully informed.[26] Moreover, he was concerned that the British might withdraw from the project and return to England after the conclusion of the European war and before the end of the Japanese war. He offered his support for postwar collaboration as a carrot in exchange for British continued support for the duration of both wars and for keeping him informed of British atomic progress. He advised Chadwick that if he was kept properly informed of British activities, he would be better able to deflect adverse criticism or suspicion of British motives that would inevitably come from

Congress. He was anxious not to give the isolationists an opportunity to destroy postwar collaboration.[27] His deception worked. On March 7, Anderson wrote to Chadwick: "I would see no objection to your following up your conversation with Groves by giving him full details of the preliminary work which has been set on foot."[28] Furthermore, Oliphant, who had been the one talking about returning to his university in England, changed his mind. This openness and flexibility on the part of the British backfired. When Groves discovered the extent of the preliminary work of the British, it helped to harden his resolve against postwar collaboration.

Another source of conflict between the British and the Americans revolved around what we will call the French Situation. The Americans had always been concerned about the number of foreigners on the British team. In particular, the French posed a special problem, since at least eight of them worked in Canada and the United States; five of them, including Halban, were on the Montreal team. Groves was concerned that they would take the atomic secret back to France after the war. Goldschmidt stated that "Groves would have liked to see us quit the project right away, but since he could neither cut out our tongues nor have us interned (a year later he did consider doing just that), he preferred, a lesser evil, to keep us sworn to secrecy in Canada indirectly under his thumb, at least until the bomb was used or the fighting stopped."[29] Halban had two strikes against him: not only was he French, but also he had developed close ties with ICI. After the signing of the Quebec Agreement, the Americans took immediate steps to have him removed from his leadership role in Canada. They insisted that an Englishman or Canadian head the Canadian laboratory. The British buckled under, as they had done with Akers, and appointed Cockcroft to replace him in April 1944; however, for the moment, Halban remained as part of the scientific team.

The appointment of Cockcroft did not end U.S. concerns about the French team. Between October 1944 and the end of January 1945, a more serious problem arose surrounding a U.S. perception that the British had breached the terms of the Quebec Agreement. What resulted was another unsettling episode in the uneasy relationship between the United States and the United Kingdom. The incident provides further illustration that atomic collaboration in wartime was no more harmonious than in the postwar period.

One may recall that in 1938 the Paris team had discovered the possibility of a chain reaction in the fission of uranium. The team had assigned its patent rights in these discoveries to the Centre Nationale de la Recherche Scientifique—a body comparable to the U.S. Office of Scientific Research and Development. Just before the fall of France, Halban and Kowarski, with the French stock of heavy water, had fled to England with the knowledge and approval of the French government. They carried with them an official letter from the minister of armament authorizing them to take the heavy water to England. In addition, Halban had been instructed to act as the "trustee of the French interests" in atomic energy matters. Subsequently, Halban entered into an agreement with the British whereby he undertook to use his best endeavors to persuade the French government to agree to assign to the British government all rights in the patents held by the French. In return the British undertook to reassign to the French government all rights in these patents for France and the French empire. Moreover, the French government was given the same territorial rights over any future patents applied for by Halban and Kowarski.[30] On the surface the contract seemed innocuous. On August 5, after it was negotiated but before it was signed, Anderson informed Bush that he had acquired all of Halban's and his group's rights in their inventions. However, he erroneously stated that the British government, in exchange, had conceded to the "inventors" patent rights for France and the French empire.[31] Bush was pleased that Anderson had taken steps to keep the control of atomic energy in the hands of the governments of Britain and the United States.[32]

For two years the contract remained buried in Anderson's file until November 1944. In May 1944, the Americans received their first piece of unsettling news. One of the French scientists, Pierre Auger, resigned to prepare himself for a return to France to "aid in the rebuilding of French science." As Auger had worked at Montreal, Groves thought that his return to France would be a violation of the project's security. London, however, did not share his concerns. Earlier, Louis Rapkine and Francis Perrin, two other French scientists, had left for Africa to join the provisional government of France on the eve of the D-Day invasion. On June 6, the liberation of France came closer to reality with the long-anticipated Allied landing on the beaches of Normandy. By October, the trickle of scientists leaving for France fast became a torrent when Jules Gueron requested permission to return to France to settle personal affairs. Groves suggested, and the British protested, that Gueron be placed under surveillance in France and that he be ordered not to meet Frédéric Joliot, whose communist sympathies were by then well known to the United States. It was then that Anderson informed the Americans for the first time that some of the French nationals working in Canada were, to quote Groves, "virtually representatives of the French Committee for National Liberation and were employed under an agreement which permitted them to return to France whenever 'their scientific position in the French Government service' made it desirable."[33] Groves also discovered that Gueron was still a French civil servant whose salary was paid not by the British but by the French government. More disclosures were yet to come.

Early in November, the torrent became a flood when Halban arrived in London; he too requested permission to go to France. Groves was up in arms. It was quite obvious to him that Halban would seek out his mentor, Joliot, and since Groves had "no confidence whatsoever" in Halban and regarded him as untrustworthy, he feared that disclosures would be made to Joliot. Anderson argued that a refusal would prod the French to raise prematurely French participation in atomic energy work. Moreover, it would create a controversy that could prevent the acquisition of the patent rights held by the French government.[34] Anderson's argument puzzled Groves, who, on the basis of Anderson's August 5 letter to Bush, was under the impression that the patents had been held by the two Frenchmen and that they had already turned their rights over to Anderson. Groves surmised that Anderson had no such rights and was angling to give secrets to the French government in exchange for the patent rights.[35]

Subsequent events did nothing to allay Groves's fears. Anderson sidestepped Groves and others "in the know" in Washington and persuaded the U.S. ambassador to London, John Gilbert Winant, with whom Anderson had worked in negotiating the Belgian Agreement, to agree to Halban's visit. Both men agreed that Halban could see Joliot and give him the "barest outline" of work on slow-neutron experiments. In view of Halban's and Joliot's past association, Anderson believed that Halban "had a moral obligation to renew contact." And if contact was renewed, he could not see how Halban could "avoid giving some account of the continuance of the work initiated in Paris."[36] Before Groves could recover from, in his view, this attempt by Anderson "to curry favor with de Gaulle and thereby increase the influence of England over the French Government,"[37] Halban returned from France to reveal that in his discussions with Joliot he had gone beyond the "barest outline." Groves was livid. He informed the secretary of war that he was suspending the flow of information to the British because they had violated the terms of the Quebec Agreement by communicating information to a third party without the consent of the president.[38]

The president was informed of the controversy. In the process Groves and Stimson portrayed Anderson as "a big man with black beady eyes" who was "dominated by 'Imperial' instinct," who was "deliberately attempting to deceive and double-cross the Ambassador and us at every possible turn" and hence should not "be trusted under any circumstances."[39] The president agreed that it was impossible to bring the French on board, but the question of suspending collaboration was never seriously considered, especially since the president was assured that Churchill was not in on the situation, only Anderson, who was "trying to run the show."

As a matter of fact, in the spring of 1944 Churchill had sharply rejected a submission by Sir Anthony Eden, the British foreign secretary, and Anderson that the French should be assured that they would be allowed to participate in the program as soon as security conditions permitted.[40] Anderson was adamant that the French should be given fair treatment after the war based on their pioneering contributions to atomic energy and their contributions to the bomb project. He assured the Americans that he had incurred no specific obligations to the French government, but he felt that he had an "obligation of honour" to persuade the United States to share with the French any benefits resulting from the slow-neutron side of the work.[41] In a meeting with Anderson in London, Joliot made the same arguments, after which he added that if France were not admitted to collaboration with Britain and the United States, she would have to turn to Russia, to whom initial enquiry already had been made.[42] Anderson took the long-term threat seriously. He gave Joliot vague assurances that after the conclusion of the war Britain would hold further discussions relating to the industrial application of atomic energy.

Groves went along with this compromise, but the United States made no commitments. He recommended that if Joliot rejected the British declaration of intention, the French scientists should "be placed in confinement in Canada and not be permitted to communicate with anyone."[43] Churchill's views were no less extreme than Groves's. He told Anderson that if there were any real danger of Joliot's going to the Russians, he "should be forcibly but comfortably detained for some months."[44] Anderson had to rush off a cable to Yalta, asking Eden to ensure that Churchill did not follow through on his detention threat. Churchill was adamant that the British and Americans guard their atomic secret. At Yalta, he had made his views emphatically clear to Roosevelt, who had "shocked" Churchill by casually raising the idea of revealing the secret to Stalin.[45] He did not have to act on his threat. Groves suggested that if Joliot were receptive to the British overture, then the contracts of Gueron, Kowarski, and Goldschmidt should be renewed, with steps taken to limit their knowledge of the project. (Gueron turned down the renewal and insisted on returning to his university post at Strasbourg.) Halban was to be retired and confined to the United States or Canada.[46] Eventually, Halban agreed to terminate all official contacts with the Montreal Laboratory by April 1, 1945, and to proceed to the United States, where he hoped to "latch on" with an American university.[47] One should add that Groves attempted at every turn to discourage any American university from hiring him. It is worth noting as well, in defense of Groves's security concerns, that after the war General Charles de Gaulle admitted that although the Americans and British had told him nothing about the atomic project, he had been briefed by the

French scientists working with the British.[48] The whole patent situation petered out after the war when the French government repudiated the Halban
agreement.

Anglo-American collaboration, however, was not destined to end with the
war. Churchill and Roosevelt saw to that. Almost all of Churchill's atomic
energy advisers had raised the issue of Anglo-American postwar policy. These
advisers initiated steps to remove the uncertainty. They held informal talks
with Bundy to explore the possibility of extending the terms of the Quebec
Agreement to include postwar collaboration. Bundy discouraged an official
approach because he did not believe that Stimson and his advisers would be
receptive.[49] Anderson then approached Cherwell, who, despite his doubt that
it could be arranged, agreed that the prime minister and the president should
take up the issue.[50]

On September 12, 1944, Cherwell advised Churchill to broach the topic
with Roosevelt so that the British would be in a position to make their own
atomic energy plans for the future. No one could have imagined the informal
atmosphere under which Churchill would raise the subject. It was not done at
the second Quebec Conference in Canada, which had just concluded, but after
both men returned to Hyde Park, the president's residence in New York.
While both men were in the garden, with no aides present, they discussed the
vital issue of postwar collaboration and use of the atomic bomb. The document to which both men agreed and signed is called the Hyde Park Aide-
Memoire. It provided that "full collaboration between the United States and
the British Government in developing Tube Alloys for military and commercial purposes should continue after the defeat of Japan unless and until terminated by joint agreement." It also rejected the idea that the "secret" should be
shared with the rest of the world. Both men agreed that when a bomb was
finally available, "it might perhaps, after mature consideration," be used
against Japan. Finally, Niels Bohr, the Danish scientist who had escaped to
England, was to be investigated, and steps were to be taken to ensure that he
would disclose no information to the Russians.[51]

The inclusion of Niels Bohr in the Aide-Memoire stands as a clear illustration of the cavalier and impulsive manner in which this document was drawn
up. Neither man's advisers shared his perception of the respected Danish scientist. Earlier, at separate meetings with Churchill in London and Roosevelt
in Washington, Bohr had argued that, in the interest of further international
peace and mutual trust, the Americans and British should inform the Russians
of the atomic project. Churchill had no inclination to do so, but Roosevelt did
lead Bohr to believe that he saw some merit in the idea. Bohr, like others
before him, was taken in by Roosevelt's gift for leaving others with an impres-

sion of a favorable audience. In fact, in the course of their discussions at Hyde Park, it was Roosevelt who informed Churchill that Bohr had made unauthorized disclosures about the bomb to Justice Frankfurter and that Bohr was in correspondence with a Russian professor with whom he had discussed atomic matters. The following day Churchill wrote Cherwell that he and the president were "much worried about Professor Bohr. . . . It seems to me Bohr ought to be confined or at any rate made to see that he is very near the edge of mortal crimes. I had not visualized any of this before [the president's disclosures], though I did not like the man when you showed him to me, with his hair all over his head. . . . I do not like it at all."[52] Cherwell and Anderson rushed to Bohr's defense, but the opinion already was formed. Revealingly, neither Bush, Conant, nor Groves regarded Bohr as a security threat. Apparently, Churchill and Roosevelt weighed "the evidence" and formed their own opinion without consulting their advisers. Bohr had discussed the bomb and its postwar implications with Frankfurter, but the justice was already aware of the existence of the bomb project. Bohr believed that steps should be taken by the United States and the United Kingdom to win Soviet trust in the postwar era by earning it in the wartime period through a disclosure. That disclosure could set the foundation for international control of atomic energy. In the absence of mutual trust among the powers, he foresaw the chances of an arms race at the end of the war.

With reference to the postwar atomic matter, was the president serious about postwar collaboration, or was the Aide-Memoire a useless spur-of-the-moment commitment that Roosevelt had no intention of honoring? Was this another attempt to appease Churchill? Sherwin believes that the president took the Aide-Memoire quite seriously. Roosevelt's acceptance of the agreement hinged on his recognition of the value of Britain as an ally in the postwar period. As Sherwin put it, "Economically and militarily secure, and armed with atomic weapons, Great Britain would be America's outpost on the European frontiers, the sentinel for the New World in the Old."[53] Did he then agree to postwar atomic collaboration because without it Britain would be powerless to act as a sort of U.S. military surrogate in Europe? One is challenged to bless this hastily drawn-up agreement with such lofty aspirations. Perhaps Churchill left Hyde Park believing that the document in his briefcase would elevate Britain to the pinnacle of atomic power. But it is questionable that Roosevelt had any such design in mind. If the president contemplated such an extensive and global mission for the United Kingdom, why did he not formally initiate discussions with Churchill? Rather, he agreed to the postwar terms of the Aide-Memoire only after Churchill, on the urging of his own advisers, brought it up. Moreover, Churchill wasted no time in fully informing Anderson and

Cherwell of what had transpired at Hyde Park. Cherwell was pleased, for the new arrangement seemed to presuppose a military alliance. Roosevelt, on the other hand, kept the existence and details of the Aide-Memoire a secret from his most trusted advisers, probably because he knew they would have taken him to task. Not a single adviser favored a policy that envisaged a partnership with the British to the exclusion of all other nations. Not even Stimson was made aware of the Aide-Memoire. He saw it for the first time on June 25, 1945—more than two months after the president's death. The British, who by then had brought the existence of the document to the attention of the Americans, submitted a photocopy to Stimson.

From the outset, Roosevelt handled the document in a dismissive manner. He handed the Aide-Memoire, this key to world security and American global power, to a clerk, instructed him to file it, and never, between then and his death seven months later, asked to see it again. Nor did he offer the clerk any further instructions pertaining to its importance or filing. Since the Aide-Memoire was entitled "Tube Alloys," the clerk concluded that the document dealt with torpedoes and, accordingly, filed it among an admiral's papers. It was not recovered until 1957. Moreover, the very title of the document, "Tube Alloys," the British code name for the atomic bomb project, seems to indicate that this was another British document, probably dictated by Churchill. One year later, Admiral William D. Leahy, who was at Hyde Park at the time, informed a conference, called by the new president Harry Truman and attended by Groves, Bush, Conant, Stimson, and Secretary of State James Byrnes, that to his knowledge no agreements had been made about sharing military secrets. He stated that Roosevelt's attitude had been that military secrets concerning the bomb should not be divulged but that the United Kingdom should share equally in any industrial benefits arising from atomic energy.[54]

Three days after Hyde Park, in a general discussion with Bush and Lord Cherwell, the president did make references to continued indefinite Anglo-American atomic collaboration and to Britain and the United States assuming the role of atomic postwar policemen.[55] Were these statements made for the benefit of Cherwell, who could be counted upon to take them back to Churchill? Was he trying to reassure Churchill that he was serious about the Aide-Memoire, which his clerk had already filed away? There are no certain answers. It is also quite revealing that the president raised these issues in the presence of Bush, whom he knew was against any such Anglo-American partnership. The possibility exists that the president, in his own inimitable way, was attempting to probe Bush for an opinion or even to generate a debate between Bush and Lord Cherwell to help him formulate his future policy.

There were few men better equipped to debate the wisdom of a postwar Anglo-American atomic policy. Either way, Bush failed to challenge the policy when the president presented him with the opportunity. Rather, he sat there feeling "very much embarrassed" and isolated from atomic diplomacy. We may never know the state of Roosevelt's mind at this time; we can only speculate that he did not take the Aide-Memoire as seriously as the British were led to believe.

Although he was not officially informed, Bush suspected that Roosevelt had made some type of commitment at Quebec. After the talk with Cherwell and the president, he shared his suspicions with Conant and Stimson. It seemed to him that Roosevelt and Churchill believed that they could establish a monopoly over atomic energy and use it to control the peace of the world. Bush disagreed with them and expressed views that showed that he had given thought to the future threat of atomic weapons.

> I felt that this extreme attitude might well lead to extraordinary efforts on the part of Russia to establish its position in the field secretly, and might lead to a clash, say 20 years from now. On the other hand, I felt that if there were complete scientific interchange on this subject among all countries that there would be much less danger of a secret race on military applications as the art changes. I also felt that there might be a possibility that an international agreement involving all countries for the control of this affair might have some hope of success, and that it most certainly ought to be explored and carefully analyzed.[56]

This was a remarkable turnaround for a man who, two years earlier, had jealously guarded American superiority in the field. By late April and early May 1944, he and Conant had started considering the international implications of atomic weapons. It was apparent to them that the United States would be unable to maintain its superiority over atomic weapons and that an arms race could result as nations scrambled to upgrade their arsenals. Between then and late September, they endorsed the creation of a postwar international atomic energy commission to regulate the spread of atomic weapons. By then the United States would have demonstrated the power of the bomb to the world, including the Soviets, with whom scientific but not technical information would be shared. After the September 22 meeting, Bush was disturbed to learn that the president and Churchill had other plans. The president, he complained, was proceeding without the advice "of the best possible minds" who had given the issue close thought. He compared the present situation with that prevailing at the time of the Quebec Agreement. The president, he commented, had signed that agreement after having been advised quite completely about the wartime aspects of it but without full advice in regard to some of the political and postwar clauses.[57] As late as June 22, 1951, Conant still displayed quite a bit of irritation over what he described as the "somewhat cavalier" manner in which Roosevelt dealt with the international atomic arrangements. He expressed the conviction that the president had no idea of the enormous importance of "our secrets." Furthermore, he did not believe that the president possessed an awareness of the consequences of his action.[58]

Roosevelt and Churchill must be held partly responsible for the subsequent problems that cropped up after the war. The wartime agreements were based too closely on the personal relationship existing between both leaders. They had a gentleman's agreement, sealed by a handshake. From the very beginning Churchill had his doubts about the fairness of the Quebec Agreement, but he signed it anyway because he trusted Roosevelt. By 1943, Roosevelt and Churchill had formed a tight friendship. It was based on mutual respect, kindness, and understanding. Both men admired the other's charm, wit, and adeptness at manipulating the press and public opinion. Initially, the friendship was cultivated by Roosevelt's support after the fall of France in the form of the Destroyer-Base deal, Lend-Lease, and the Atlantic Charter. It grew after the United States entered the war, as both men, in times of travail, motivated each other with exchanges of poetry and biblical verses. They differed on the value of British colonialism and its effect on the natives, the invasion of Europe, their perceptions of Stalin, and the Soviet's role in postwar Europe, among other issues, but their friendship was never seriously threatened. Churchill anticipated his summits with Roosevelt not only as occasions to discuss matters of state but also as an opportunity to escape from the pressure of leadership in London. They were mini-vacations with a president who put him at ease and encouraged him to relax. Roosevelt never lost sight of, or concealed, the fact that Churchill was closer to the line of fire than he was or that the English people, more than the Americans, lived under the daily threat of German conquest.

Churchill and his aides should have discussed with Roosevelt and his people the binding nature of the Hyde Park Aide-Memoire and the extent to which a new president would be bound by what was no more than an executive agreement with a shelf life of the contracting president. Roosevelt's aides had an excuse for not debating the ramifications of the Aide-Memoire—they knew nothing about it. But Churchill's advisers had no excuse. Neither Roosevelt nor Churchill had paused to consider the possible effects of politics and fate on the ever-changing state of international affairs and the national interest. Roosevelt died on April 12, 1945, and the British electorate voted Churchill out of office three months later, leaving behind a vaguely defined agreement for postwar collaboration, the existence of which no one in the United States knew about. When one considers the effort expended by those who were left behind in the United States to avoid wartime collaboration with the British, one could only anticipate further problems in the future. The men who had done the most stonewalling were still in positions of authority or consultation, and the new president, Harry Truman, had to turn to them for atomic advice in the postwar era.

FOUR

Immediate Postwar
Cooperation Comes to a Halt

President Roosevelt died on April 12, 1945, and Vice President Harry Truman succeeded him as president of the United States. Three months later, the British electorate voted Prime Minister Churchill out of office and replaced him with the deputy prime minister of his wartime coalition government, Clement Attlee of the Labour Party. Although the two new leaders had been highly placed in their respective governments, neither man knew anything about the Manhattan Project. Fate could not have selected a more uninformed pair of world leaders to address the critical issues surrounding the atomic bomb in the waning year of the war and the immediate postwar period. Both Roosevelt and Churchill had deliberately chosen not to keep their deputies abreast of atomic development. Thus, both men were forced to rely, for better or worse, on the expertise of wartime players from the previous administrations for atomic energy information. The two new leaders also blindly accepted an existing policy, the far-reaching significance of which they barely understood. Both men rushed feverishly into an understanding to continue collaboration and would have done so if General Groves had not taken upon himself the responsibility of warning off his president. Groves's warning brought collaboration to a halt in mid-1946.

The fact that Truman was not kept informed about atomic energy matters should not be interpreted, by itself, as a sign of the president's lack of trust in his vice president. We have already noted that Roosevelt often kept even his close atomic energy advisers uninformed about his private discussions with Churchill. James Byrnes, director of war mobilization, and soon to become Truman's secretary of state, was not told of the Manhattan Project until the summer of 1943.[1] However, this knowledge did not translate into a flow of atomic information. As late as March 3, 1945, long after Groves had assured the president of the imminent success of the project, Byrnes had suggested to

Roosevelt that a committee of independent scientists be appointed to investigate and review the success or failure potential of the project. He added in the memo that he knew "little of the project except that it is supported by eminent scientists."[2] (Roosevelt came very close to selecting Byrnes as his running mate in 1944, in which case Byrnes and not Truman would have succeeded him as president.) As for Secretary of War Stimson, even though he was involved from the inception of the project, he did not see a copy of the Hyde Park Aide-Memoire until the British sent him a photocopy on July 18, 1945.[3] The State Department knew nothing about the research activities in progress until it was informed just before the Yalta Conference of February, 1945.[4] Moreover, Groves, in the interest of security, chose not to involve the department in his negotiations for uranium. Joseph Volpe reported that when he told Dean Acheson, undersecretary of state, in early 1946 of the various commitments entered into by Groves and his people, Acheson was "absolutely flabbergasted." His comment was "Why, you sons of bitches set up your own state department."[5] The vice president was in a good company of uninformed officials.

Unknowingly, Truman came perilously close to blowing the secrecy lid off the project while he was still a senator. Up until 1944, the administration financed the atomic project with funds drawn from several defense appropriations, but by February 1944, as the project began eating up funds, it decided to approach the congressional leaders of both Houses for direct appropriations. The congressional leaders piloted the appropriations through the Houses without public debate. However, as a member of the Appropriations Committee, and not privy to inside information, Senator Truman raised a red flag when he noticed that the appropriation request did not disclose the nature of the project. Since he was also chairman of the Senate special committee to investigate the National Defense Program—the Truman Committee—he dispatched his investigators to snoop around the nuclear facilities at Oak Ridge, Tennessee, and Hanford, Washington. Stimson quickly arranged a meeting with Truman and, without disclosing the nature of the project, warned him off on the grounds that it involved a top secret of the government. Truman obediently called off his investigators.[6]

Not until two weeks into his presidency did he learn what that secret project was all about. On April 25, Stimson, joined later by Groves, briefed him on the scope of the operation. He told him that the weapon would be completed in four months and that the British had contributed to its development. Truman was also informed that both countries temporarily controlled the scientific and industrial knowledge and raw materials required for a bomb but could not retain that position indefinitely, since scientists from other countries were acquainted with segments of the processes. Stimson even broached the subject of the difficulties surrounding the challenges of international control and emphasized that it was a problem that would have to be addressed in the future. He also proposed, and Truman approved, the appointment of an interim committee[7] to make policy recommendations to the president.[8] By the end of April, Truman was fully apprised of the Manhattan Project.

Clement Attlee was still in the dark and would remain there until July, when his election victory over Churchill thrust him, as totally unprepared as Truman was, into the light of atomic revelations. After graduating from University College, Oxford, the twenty-two-year-old Attlee practiced law, involved himself in social work, and later taught social science at the London School of Economics. During World War I, he served as a major in the army, and after the war ended, he became actively involved in politics. He was one of the few Labour ministers to retain his seat in the 1931 elections. He became leader of the Labour Party in 1935. Two days after Hitler reoccupied the Rhineland, on March 7, 1936, Attlee's Labour Party opposed the Conservative government's defense white paper to increase the strength of the army, navy, and air force because he considered the move provocative. By 1938, however, he agreed with Churchill that the Nazis were a menace. In September of that year, his party was prepared to join Churchill and his handful of friends in a rebellion against Prime Minister Chamberlain if he returned from Munich with a dishonorable deal. Both he and Churchill were apprehensive over the Munich Agreement. Attlee called it a "humiliation" and a "victory for brute force,"[9] but Churchill was not willing to join him in a rebellion. The rebellion came two years later, when Chamberlain offered to form a coalition government. Attlee and the Labour Party refused to serve under him. They and the majority of Parliament rallied around Churchill. Within a year, Churchill appointed Attlee deputy prime minister. The coalition government ended with the defeat of Hitler. At its annual conference at Blackpool on May 21, 1945, the Labour Party voted for a return to party politics. It had a strong desire to introduce a socialist agenda, which Churchill resisted on financial and ideological grounds. The electorate, despite Churchill's warning that Labour would introduce totalitarianism and a "form of Gestapo," cast its vote for social security, full employment, free housing, and a host of other promised social benefits. Attlee won the election and walked into the unknown world of atomic energy.

Churchill was even more obsessed with secrecy than Roosevelt was. Only seven members of the wartime coalition government knew anything about the atomic project, and not one of them was a member of the Labour Party. Sir Stafford Cripps, minister of aircraft production and a Labourite, got wind of the project and attempted to inform himself, but Churchill's people quickly warned him off.[10] Even the Chiefs of Staff were not fully briefed. As late as January 1945, Anderson pleaded for Churchill's authority to put them "fully into the picture," on the grounds that their counterparts in the United States' army, air force, and navy were in the know.[11] Churchill grudgingly agreed. The most illustrative example of Churchill's secrecy can be found in his treatment of Sir Henry Tizard. In early January he approved the appointment of Sir Henry to investigate the future potentiality of weapons of war over the next ten years. Yet the prime minister refused to provide him with research and development data about atomic energy. The representations of General Ismay, chairman of the Chiefs of Staff, Anderson, and a personal minute from Sir Henry himself[12] could not sway Churchill. He cut off all arguments with a blunt refusal and an expression of his annoyance. "He [Sir Henry] surely has lots of things to get on with without plunging into this exceptionally secret matter."[13] The report was obsolete even before it was written.[14]

It was with a sense of shock that the new Labour government absorbed atomic energy briefings after it came to power in July. With reference to the Quebec Agreement, Hugh Dalton, chancellor of the Exchequer, classified it as an "astounding sell-out" that gave the United States the power to deny the British any industrial rights. He was peeved that the coalition cabinet had not been consulted and believed that the inevitable publication of the agreement would destroy Churchill's reputation.[15] The new government did not have the liberty of submerging itself in self-pity or anger. There were pressing atomic energy matters to be addressed, and there was no qualified Labourite to provide direction. Hence, three days before the United States dropped the atomic bomb on Hiroshima, Attlee accepted the secretary of the cabinet's suggestion that Sir John Anderson, who had retained his seat as an Independent for the Scottish universities, be kept on to head a new Advisory Committee on Atomic Energy.[16] For the next year or so, this committee played a leading role in getting the British atomic program off the ground. Also, it bought time for the Labour ministers to understand the issues, after which the committee lost its importance. During this transitional period, the prime minister immersed himself into atomic energy matters; he made all major policy decisions with the advice of a few confidants. Like Churchill before him, but for different reasons, as we will discuss later, Attlee shrouded the British atomic program in secrecy. Most cabinet ministers were excluded from deliberations. The insiders were Herbert Morrison, lord president of the Council; Ernest Bevin, foreign secretary; Stafford Cripps, now president of the Board of Trade; Arthur Greenwood, lord privy seal; Hugh Dalton; and, after the project was placed under the Ministry of Supply, John Wilmot. These members met under the umbrella of an ad hoc committee, called "Gen 75," that was convened in August 1945.[17]

Collaboration was among the pressing atomic issues that confronted Attlee. The outgoing government had briefed him on the American partnership and shown him a copy of the Hyde Park Aide-Memoire. Any misgivings he might have had about Churchill's handling of the whole matter were soothed by the knowledge that the Americans were willing to cooperate with the British in the future. In the weeks and months following the demonstration of the atomic bomb's explosive power in Japan, all public statements coming out of Washington tended to portray the British as an equal partner in the development of the bomb. Attlee developed this theme. As early as August 8, Attlee wrote Truman a letter that left no doubt that he perceived the British as an equal partner in the development and control of the new weapon. "I consider, therefore, that you and I, as heads of the Governments which have control of this great force, should, without delay, make a joint declaration of our intentions to utilize the existence of this great power, not for our own ends, but as trustees for humanity."[18] Eight days later, he rushed off another letter reminding Truman of the close collaboration that had existed between both countries during the war. This cooperation had led to the development of new and improved weapons and techniques and was made possible by the "frank exchange of information and personnel." He expressed the hope that both men would issue directives authorizing continued collaboration.[19] He could have added also that both countries had arrived at a "mutual consent" before the atomic bomb was used against the Japanese.

It appears that at this formative stage of his atomic policy Truman too favored collaboration. His advisers had already briefed him on the British contribution to the project. On October 8, at a press and radio conference at Reelfoot Lake in Tennessee, he informed his listeners that he was not averse to sharing the scientific knowledge that had resulted in the development of the bomb because it was already known to scientists worldwide. However, he explained, the practical "know-how" would not be shared. In response to another question, he added that Great Britain and Canada were already in on the "know-how" and were partners with the United States. He explained that the project had been transferred to the United States from Britain during the war because the United States was better able to meet the required expenditures.[20] Again, on October 31, he repeated that Great Britain knew as much about producing the bomb as the United States did.[21] On the surface it would seem that the most uncertain question floating around London and Washington was not whether information should be shared with the United Kingdom—the public perception was that the United Kingdom already had the know-how—but whether Russia should be made a full member of the atomic club. This issue was hotly debated on both sides of the Atlantic inside and outside of the respective cabinets, as we shall discuss in the next chapter.

Despite the public acknowledgment of the British role, already an undercurrent of opposition had started to build at the highest level of the president's cabinet. On October 10, Secretary of State James Byrnes, Secretary of War Robert Patterson, who had recently replaced Stimson, and Secretary of the Navy James Forrestal held a meeting in which they agreed that the British apparently intended to influence U.S. atomic policy through the Combined Policy Committee (CPC). To guard against this intrusion, the three secretaries discussed the possibility of disbanding the committee. George Harrison, Patterson's aide, suggested that in the interim it was important to have strong American representation on the committee. To this end they agreed that General Groves was the best man for the job (Attlee had just proposed Roger Makins[22] as the United Kingdom's) and that Byrnes would replace Dr. Conant.[23] Thus, behind the scenes, the three secretaries already had started shoring up their defenses against the British.

On September 25 Attlee again wrote Truman requesting that both leaders get together to discuss their plans for handling the bomb. He pointed out that the manufacturing and raw material monopoly that both countries shared could not last.[24] Hence, both powers should form a consensus to place all atomic weapons under the control of the United Nations.[25] The president's thoughts were fully tuned to those of the prime minister. Already he had started the process of building a consensus for a policy on atomic energy. Acheson stated that five days before receiving Attlee's letter, Truman had already decided to raise not only the subject of the national control of atomic energy but also the subject of international control in his upcoming address to Congress. Senate Majority Leader Alben Barkley and House Speaker Sam Rayburn had assured the president that the inclusion of international control would strengthen his address to Congress.[26] On September 21, Truman held the first of several luncheons to elicit the opinion of his cabinet. Though contentious, these meetings helped the president to formulate his atomic energy policy.[27]

Attlee's timely telegram served to reinforce the policy for which Truman was seeking a consensus. By no means did Attlee force upon the president the ideas of a summit and international control. Eight days after receiving Attlee's letter, he addressed a message to Congress recommending the creation of an Atomic Energy Commission. He revealed also that he was about to initiate discussions with Great Britain and Canada to effect an agreement for international control.[28] Two days later, on October 5, he responded to Attlee's letter, indicating his willingness to meet with the prime minister.[29] Unfortunately, somewhere between transmission from Truman to Attlee the reply was misplaced. Not until Attlee wrote an urgent follow-up expressing his concerns over not hearing from the president did he receive a copy of the reply on October 13.[30] Four days later, telegrams were exchanged in which Truman agreed to meet Attlee and Mackenzie King, the prime minister of Canada, in Washington anytime between the fifth and fourteenth of November.[31]

Both Attlee and Truman were in agreement that international control should be examined at the summit. However, their subordinates also recognized that the moment of the summit could be seized to address other collaborative issues. In particular, the British saw the need to reexamine the terms of the Quebec Agreement in light of the prevailing peacetime conditions and in light of indubitable signals that the American members of the CPC regarded the agreement as having lapsed with the end of the war. By November 3, Sir James Chadwick had complained that since the end of hostilities the United Kingdom had received no information from the CPC.[32] It was clear to the British that the prevailing view among Truman's people was that the Quebec Agreement was an executive agreement drawn up by a president who had exercised his emergency powers to meet the demands of an emergency situation. Thus, Attlee's Cabinet Advisory Committee on Atomic Energy believed that if they were to preserve the collaboration they had enjoyed under the agreement, even within the framework of some eventual international plan, they must themselves take the initiative in proposing a new agreement. The Advisory Committee saw the need to act quickly before the views of the new administration became crystallized and while the machinery of the CPC and of the Combined Development Trust (CDT; the raw material arm) was still active.[33] More precisely, the Advisory Committee had serious reservations about the "sell-out" clause of the agreement, article 4. The article imposed a restriction upon the United Kingdom's freedom to engage in the commercial exploitation of atomic energy without the consent of the U.S. president. This clause undermined the sovereignty of the British and made them susceptible to the whims of the American leaders. Additionally, the British wanted to clarify the final provision of the agreement, limiting the conditions governing the exchange of information. In its present form, it did not provide a satisfactory basis for future collaboration.[34] The CPC reserved the right to determine when and whether atomic information should be exchanged. Even the provisions with regard to allocation of raw materials needed to be redefined. They stated that material should be allocated in accordance with the requirements of the atomic programs agreed to by the CPC. The practical effect of that stipulation was that during the war all available material had gone to the

United States. With the end of the war, the British anticipated that they would be engaged in some form of atomic research—if not bomb production, then at least peaceful application of atomic energy. Hence, changes were necessary to facilitate a more equitable distribution of raw material. Thus, although the summit was inspired by the need to agree upon common international policy on atomic energy, the revision of the Quebec Agreement or its substitution by another document was "one of the principal subjects" that the prime minister and Sir John Anderson were prepared to handle in Washington.[35]

Other factors quickly surfaced to add urgency to the reexamination of the agreement. In the weeks leading up to the summit, the issue of collaboration and the United Kingdom's own atomic energy plans were raised in Parliament time and again by opposition members.[36] On September 28, Attlee was forced to seek the help of Winston Churchill, now leader of the opposition, in keeping questions of atomic raw material off the Order Paper and thus off the floor.[37] Churchill agreed to cooperate for fear of jeopardizing the secret raw material agreements with Belgium, Holland, and Brazil.[38] This understanding, however, did not succeed in restraining Captain Raymond Blackburn, the Labour representative for King's Norton in Birmingham, who caused a sensation in the House on October 31 by making the first public reference to the Quebec Agreement and article 4. Not only were the Labour backbenches crowded, but as he spoke, ministers rushed in to pack the front benches.[39] Captain Blackburn appealed for the publication of the agreement. The government dared not agree due to the embarrassing article. Truman was not inclined to agree anyway because the Americans too had reasons to anticipate an uproar in Congress if the terms of the agreement were disclosed—particularly article 2, which stipulated that the United States would not use the atomic bomb without the consent of the United Kingdom. This date marked the beginning of a tacit understanding between both administrations not to publish the text of the agreement. When President Truman was asked the following day about the agreement, he denied its existence.[40] The commotion in Parliament underlined the need to do something about the existing agreement.

In preparation for the summit, Attlee sought the views of the Chiefs of Staff, the Washington-based Joint Staff Mission (the military wing of the British embassy), the Anderson Committee, the Foreign Office, and Gen 75—the Ministerial Committee on Atomic Energy. The British took the summit and the issues involved quite seriously. The same cannot be said about the American approach to the meeting. Secretary of State Byrnes went about the organization of the summit in a rather haphazard manner. It seems that he was not too keen on the timing of the summit, which threatened to adversely affect other international issues. He was concerned that the public announcement of an atomic summit would create the suspicion by the Soviets that the West was ganging up against them. This outlook could harden Stalin's resistance to modify his position and cooperate with the Far East Advisory Commission. The commission had agreed in late September to create a centralized administration in Japan under the supervision of General Douglas MacArthur. Moreover, since the United States and United Kingdom were actively engaged in the final stages of negotiations concerning the appointment of an Anglo-American Committee to investigate the Palestinian-Jewish problem, he was more anxious to clear up that issue than to plunge into a summit.[41] Thus, little was done to prepare for the summit. Although the summit was a State Department matter, Secretary of War Patterson saw the need to send Byrnes a memo on November 1, ten days before the summit. He urged him strongly to prepare for Attlee's visit by undertaking "a thorough examination of the international phases of atomic energy . . . and the problem of the war-time Quebec Agreement."[42] Byrnes was noncommittal. But the following day Patterson took it upon himself to hold discussions with Bush and to order Lieutenant Gordon Arneson to draw up a tentative set of proposals for discussions.

The British were totally aware of the Americans' unpreparedness. On November 5, Groves confided to Roger Makins, the U.K. member of the CPC, that he had not been consulted and that he thought that the Americans would be badly prepared.[43] The Earl of Halifax, the British ambassador, reported three days later the "remarkable fact" that none of the Americans hitherto concerned with atomic energy had been consulted by the president or Byrnes. "Nothing has been said to Patterson, [Dean] Acheson, [George] Harrison or Groves, nor, as far as they know, have procedures been worked out for conducting the talks. Even Bush is apparently in the dark." He ended with the prophetic statement: "This does not augur well for clear and effective treatment of the subject on United States side."[44]

Eventually, Byrnes asked Bush and Groves to submit memoranda on their thoughts, but they were not included in the negotiations until these were well underway. Two days after the summit started, Byrnes "startled" Bush by informing him that the conference had already come to an agreement on the international issue. It turned out that Truman, Attlee, Byrnes, and Anderson had made most of the decisions while on a boat cruise along the Potomac. Byrnes apologized that if he had known that Anderson would accompany them, he would have taken along Bush and Groves. Incredibly, over Bush's protest, Byrnes asked him to submit a draft communiqué on the results of the conference that he had not attended.[45] Even more remarkably, Bush's draft formed the basis of the final communiqué on the international question. Bush would comment that the general impression was "certainly that of a conference which has been somewhat chaotic due to lack of preparation and lack of organization in carrying it on."[46]

Byrnes showed little appreciation for the importance of the summit and involved himself only peripherally. Patterson and Groves would step into this vacuum, and Groves, in particular, would leave his indelible mark on what came out of the atomic discussions. On November 14, Byrnes turned over the revision of the Quebec Agreement to Patterson. Fortunately, Patterson was somewhat prepared. He had already initiated a study of the agreement, a study he thought would not be needed by Byrnes. The following day, he, Groves, and Harrison got together with a British team consisting of Anderson, Makins, Field Marshal Sir Henry Maitland-Wilson, Dennis Rickett, and Malcolm MacDonald. It was this group, with little or no input from the State Department, that worked out terms for the revision of the Quebec Agreement. It agreed not to disband the CPC and the CDT. Rather, the CPC would be responsible for drafting the appropriate terms of the new agreement. Groves and Anderson would assist the CPC in arriving at its recommendations by preparing a Memorandum of Intention listing the various points to be covered in the new document. Moreover, the consensus was that the president and the two prime ministers should issue a short statement indicating that atomic energy cooperation would continue among the three countries.[47]

Groves's influence was telling at these negotiations. It was he who introduced the quid pro quo trade-off, which conceded to the United Kingdom some collaboration and the elimination of article 4. In exchange, the United Kingdom agreed to turn over all uranium and thorium ores located in the British Commonwealth, including South Africa, to the control of the CDT for allocation in accordance with need. Since the U.S. need was far greater than that of the United Kingdom, Anderson, by accepting the trade-off, had in effect agreed to turn over all Congo uranium to the Americans. In addition, as a further concession to the United States, he agreed to the substitution of "prior consultation" as opposed to "mutual consent" for the future use of the atomic bomb against third parties. On the morning of November 16, a short statement was drawn up incorporating the changes for the two leaders to sign. Anderson insisted that the words "full and" be inserted before the phrase "effective cooperation in the field of atomic energy." Groves strongly objected to this change, but Patterson overruled him with the explanation that it made no material change in the meaning of the phrase.[48] Patterson took the short statement to the White House, where Truman and Attlee signed it.

After Patterson left, work began on the longer Memorandum of Intention. It was then that Groves staged his coup. He persuaded Anderson to agree to the inclusion of the statement "in the field of basic scientific research" after the preceding statement "there shall be full and effective cooperation." Groves would later claim that only the British minister Dennis Rickets put up "the bitterest of opposition."[49] Anderson overruled his assistant and accepted a watered-down memorandum. It included yet another restriction that "in the field of development, design, construction, and operation of plants such cooperation, recognized as desirable in principle, shall be regulated by such ad hoc arrangements as may be approved from time to time by the Combined Policy Committees as mutually advantageous."[50] Groves was well aware that the phrase "full and effective cooperation" could pose a problem for the United States in the future in terms of the scope of collaboration. With this significance in mind, he skillfully had the statement of clarification inserted. One week later, he brought the inconsistency between the language in the short statement and the memorandum to the attention of Patterson. "The last minute insertion of the word 'full' before the words 'effective cooperation' has to some extent increased the scope of cooperation beyond that contemplated in our discussions. . . . [However,] you will note that the fifth paragraph of the memorandum does not recommend 'full and effective cooperation' beyond the field of basic scientific research."[51] Within the next few months, this inconsistency would be resurrected to haunt the British hopes for collaboration.

The British had hoped to replace the Quebec Agreement with a document that could be published and submitted to Parliament and Congress. But as they left Washington on November 16, they were not too optimistic about achieving that goal. They returned to London with vague, unpublished promises of future American cooperation. As Kenneth Harris, one of Attlee's biographers, puts it, Truman explained to Attlee that because the clauses of the Quebec Agreement pertaining to the exchange of information had been secret, it would be politically impossible to divulge them and announce a new commitment to replace them. He also assured Attlee that "it should suffice for the British government to know that information about atomic energy would in fact be exchanged, but that the public announcement of 'cooperation' in the field would be put in general terms."[52] Attlee agreed to keep the full extent of the projected cooperation secret, "believing that what would not be made public would be honored."[53] Attlee left Washington believing that the language of their public announcement was left vague for U.S. domestic political purposes but that the true spirit of their agreement could be read from the minutes of the talks that were held. It would not take him long to realize that the deliberate vagueness of the announcement, engineered by Groves, was designed not to deceive the American public but to hoodwink the British delegation. Moreover, the minutes of their talks would support, not the British argument, but that of the Americans.

The British approached those discussions with a naiveté that would have been appropriate to a world governed by the chivalric code of a bygone age but not to the diplomatic arena of the twentieth century, where the primacy of national security and national interest took precedence over questions of principle. It was a naiveté based on the understanding that a gentleman could be taken at his word. The British experience at Munich, not too far removed from memory, had not served to illustrate that a new breed of "gentleman" was now engaged in the once aristocratic art of diplomacy. Time and time again over the next few years, as we shall see, the British would place their trust on the vague words of the Americans and be disappointed. They even attempted to rationalize the Americans' behavior.

Anderson's own Advisory Committee on Atomic Energy was not pleased with the Memorandum of Intention. Its members saw the ambiguity of the memorandum and were concerned that the Americans and the British might not understand a final agreement in the same sense. From the committee members' point of view, the provisions relating to the pooling and allocation of raw materials were clear and explicit, while those relating to exchange of information were too general. In the case of scientific research, they expected such cooperation to include exchanges of personnel and loans of material, such as separated U-235 and plutonium. As far as work on large-scale plants was concerned, they hoped to get full assistance in building British piles. Over the field as a whole, they wished to receive all information having any bearing on the military application of the project. To address these concerns, they recommended the avoidance of any sharp distinction between basic scientific research and work on large-scale plants. Furthermore, they agreed that they could not accept words such as those at the end of paragraph 5 of the memorandum, which stated that the exchanges of information should be "mutually advantageous." The committee proposed that if the Americans failed to cooperate fully on the exchange of technical information, the raw material provisions should be redefined.[54] This would mark one of the rare instances where the British would consider using raw materials to force the resolution of a collaboration issue. Unfortunately, from their perspective, this powerful weapon, as we shall discuss, was rarely ever used.

On that same day, the CPC met in Washington and appointed a subcommittee consisting of Groves, Makins, and Lester Pearson, the Canadian ambassador, to prepare the new document that was to supersede the Quebec Agreement. The subcommittee met on and off over the next several weeks and prepared a draft of a new agreement for consideration of the CPC. The draft report followed the recommendations of the Memorandum of Intention, but, to Groves's dismay, it omitted those provisions that Anderson's committee had found ambiguous: for example, "mutually advantageous" and "basic scientific research." It appeared that the Canadian and the British members of the subcommittee had outvoted Groves. He would later state:

> This draft was not in accord with my views as to what was fair to the US. It had become evident to me during the preparation of the draft that the British representatives were not willing to carry out the initial spirit of the superseded document as understood by me in two respects: first, they desired that the project be carried out in both countries not in the common interest, but rather in the preferential interest of the UK; and second, they desired that the UK be given detailed scientific and technical information and assistance by the US not to strengthen the joint position but rather that of the UK. For these reasons I personally was very much opposed to these documents and agreed to their presentation only to avoid the creation of resentments adverse to future negotiations.[55]

Not only did Groves agree to the presentation of the document to the CPC, but, according to Makins, he actually rewrote the paragraph, omitting the offensive "basic scientific research" phrase. Makins had suggested to him the submission of alternative texts, but Groves expressed his preference for one accepted text. Makins later pointed out to London the tactical advantage of accepting a draft based upon a proposal by Groves: "He will be committed to make it work."[56]

Did Groves accept these changes to avoid muddying the waters of future negotiations? Or did he accept them to strengthen the United Kingdom's position at the expense of the United States and so jolt Byrnes, Patterson, and the president out of their indifferent approach to the whole collaboration issue? By mid-February, Lord Halifax was commenting to London that he doubted whether Byrnes or Patterson had yet given serious thought to the problems surrounding cooperation.[57] David Lilienthal, then Director of the Tennessee Valley Authority, later to be appointed first chairman of the U.S. Atomic Energy Commission, commented in his journal that in mid-January Dean Acheson, undersecretary of state, had confided to him that "those charged with foreign policy—the secretary of state and the President—did not have either the facts nor an understanding of what was involved in the atomic energy issue, the most serious cloud hanging over the world. Commitments, on paper and in communiqués have been made and are being made . . . without a knowledge of what the hell it is all about—literally."[58] If Groves's plan was to shake them up, it did. Acheson told Lilienthal that in mid-February Groves came to see him, "a very scared man. 'This is the mess we are in,' Groves said, quite upset, 'you have to get us out of it.'"[59] Groves then voiced his concerns about the binding nature of the report that the subcommittee was about to conclude. Groves was not finished; he was willing to take no chances. Two days before the subcommittee was due to present its report to the CPC, Groves sent a copy to Byrnes, accompanied by a long letter of analysis. It is worth examining the content of this letter because no other single document played so crucial a role in influencing the direction of Anglo-American cooperation.

Groves explained that "the scope of the new agreement extends far beyond that even contemplated by the Quebec Agreement, and, in effect, will constitute an outright alliance, which can only be terminated by mutual consent."[60] He reminded the secretary that the Quebec Agreement was an executive agreement made under the president's power to conduct war and that it was fully enforced by both sides during the war but that all commitment had ended after the war. The memorandum of November between Attlee, Truman, and Mackenzie King, he interpreted, did intend cooperation in the field of atomic energy but did not "expressly or by implication provide for any joint venture of the three governments," as the new agreement suggested (1206). To avoid confusion over interpretation of the memorandum, he suggested, it was extremely important for the United States to arrive at a definition of "full and effective interchange of information" (1207). The British interpretation, he explained, called for the United States to turn over "practically all of the processing techniques and plant designs and specifications of the entire Manhattan Project, less the gas diffusion methods [which were at an advanced stage in Britain], and even some of the personnel" (1207). The United States, he argued, owed no great debt to the United Kingdom for their contribution to the Manhattan Project because the part they had played was unimportant. Moreover, he continued, Congress and the American people would view the agreement with displeasure, and the world community would doubt the sincerity of the United States in fostering international cooperation and world peace if it became known that it was taking steps to proliferate atomic weapons. Finally, he added, under Chapter XVI of the UN Charter, all treaties and international agreements had to be registered with the United Nations (1207). Thus, the agreement could not be kept secret even if both sides agreed to keep it secret. (The British on one side and the Americans on the other had already accepted that any new document might have to take the form of a secret executive agreement.)

Groves had pulled no punches. He threw every argument he could muster at Byrnes. His letter raised several critical issues at a time when postwar collaboration was at a vital crossroads. The letter would turn out to be the decisive factor that set in motion a cautionary reappraisal of postwar collaboration. It forced the president to hesitate and take another look at a policy he had inherited and embraced. Over the next several months, administration officials from the president on down would repeat Groves's argument as the basis to resist collaboration.

When the CPC met two days later, Byrnes zeroed in on the UN argument by immediately referring the members to article 102 of the charter, which stated that any such agreement would have to be registered with the secretariat of the United Nations. He reminded them that since the last meeting of the CPC, the United Nations had set up an Atomic Energy Commission to investigate the international control of atomic energy and that the establishment of that body had changed the position of the memorandum. In addition, it was virtually impossible to keep the agreement secret because he believed that the president had assumed an obligation to keep the McMahon Committee, which was investigating the domestic control of atomic energy, informed of any arrangements regarding atomic energy.[61]

The British did not interpret the Americans' fears as "a convenient legal pretext for abrogating the terms of the agreement."[62] They were well aware of the UN issue. Back in November 1945, before the prime minister's party proceeded to Washington, British Foreign Office officials had discussed among themselves the potential problem with the United Nations. It seemed inevitable to them that the United Nations eventually would handle the subject of atomic energy and that the disclosure of any secret arrangement could prove embarrassing. The official minutes quoted the relevant article 102 as follows: "Every treaty and every international agreement entered into by any member of the United Nations after the present Charter comes into force shall as soon as possible be registered with the Secretariat and published by it."[63] In addition, the British had also discovered that the fourth article of the Atlantic Charter had pledged the signatories to give all states equal access to the raw materials of the world.[64] Since the CDT had already abrogated the terms of the charter by engaging in a monopolistic control of atomic raw materials and thus was open to international criticism, the British elected to go with a secret revision of the Quebec Agreement to avoid exposure and criticism. This decision was made after weeks of debate. There is no evidence that they had taken their concerns to the Americans or that both sides had discussed, in an official forum, how best to skirt the article. One can only wonder whether they had raised the issue informally with Groves, thereby providing him with his ammunition. Like the British, Patterson and Byrnes also had stated their preference for a confidential agreement,[65] not because of UN concerns but to avoid revealing to the American public the existence of yet another secret wartime agreement drawn up by Roosevelt. (Several of them had surfaced after the war.) In the end they had all agreed on the secret executive agreement that the Americans were now saying was unworkable.

The British, then, were not caught totally off guard and accepted the basis of the American fears. As Lord Halifax put it to Anderson, "My only immediate observation is that I think difficulties which Americans feel . . . are not imaginary."[66] However, he believed that Byrnes's real preoccupation was not with article 102, but with the Senate, which was debating the domestic control of U.S. atomic energy. Under the circumstances he did not consider it likely that the Americans would conclude a secret agreement; thus, he urged London to agree to publication and to urge the Americans to accept a public agreement.[67] The prime minister was not receptive because "the difficulties in the way of publishing now a new formal agreement are insuperable."[68] He countered with another approach that was discussed by the CPC on April 15: rather than introducing a new agreement that had to be ratified by the United Nations, keeping the Quebec Agreement in place with the introduction of major amendments addressing the recommendations of the subcommittee.[69] It was an ingenious solution, but both the U.S. and Canadian delegates rejected it on the grounds that the amendments would in fact constitute a new agreement.[70] After debating the issue back and forth, Byrnes recommended that the matter be referred back to the three leaders for a ruling. Lord Halifax agreed to report the matter to Attlee with an ominous note of pessimism: the course of the discussion had left him with an uneasy feeling that the participants were likely to impair the background of collaboration. He thought that his report of the meeting "would be found gravely disturbing in London."[71] He was prescient. What followed next was a series of cables between Attlee and Truman. The end of these exchanges would leave no doubt that the letter and intent of the Quebec Agreement and Hyde Park Aide-Memoire were dead. Since these cables were so pivotal, they should be analyzed in some detail.

The very next day Attlee addressed a cable to Truman in which he stated that he was in fact "gravely disturbed" over the turn of events at the CPC discussions. He continued: "I feel that unless you and we and the Canadian Government can reach a satisfactory working basis of cooperation at least to cover the period until we see the outcome of the discussion in the United Nation Commission on Atomic Energy, we in this country shall be placed in a position which, I am sure you will agree, is inconsistent with the document."[72] Attlee interpreted "full and effective cooperation in the field of atomic energy" to mean nothing less than full exchange of information and a fair

division of the raw materials. He explained that "the wartime arrangements
. . . naturally meant that technological and engineering information has accumulated in your hands, and if there is to be full and effective cooperation
between us it seems essential that this information should be shared."[73]

President Truman was quick to respond. On April 20, he wrote Attlee that
he would not have signed the short statement of November 16, 1945, if he had
been informed that it provided for the United States "to furnish the engineering and operation assistance necessary for the construction of another
atomic energy plant."[74] He pointed out that under the Quebec Agreement the
United States was not obligated to assist the United Kingdom in the design,
construction, and operation of an atomic plant and that it was not his intention to assume that obligation. On the contrary, he argued, the working paper
prepared by Anderson and Groves showed conclusively that even in the minds
of the "gentlemen" who had drawn up the statement "full and effective cooperation" was not intended to require the giving of information for the construction of atomic plants. Rather, the working paper stated that "there shall
be full and effective cooperation in the field of basic scientific research among
the three countries."[75] (Anderson's faux pas had resurrected itself.) Other areas
of cooperation, he pointed out, were to be regulated by the CPC on a "mutually advantageous basis." Truman ended by rehashing two of Groves's arguments against full cooperation: the international control issue and U.S.
public opinion.

It took Attlee seven weeks to respond, and when he did it was in a lengthy
telegram in which he did not mince words nor conceal his displeasure behind
the nuances of diplomatic language. He traced U.S.-U.K. atomic relations
from 1941, claiming that from the beginning the British had had the resources
and scientific and technical skills to develop their own atomic project. However, after President Roosevelt had initiated discussions on a joint project, the
United Kingdom had agreed "in the confident belief that the experience and
knowledge gained in America would be made freely available to us, just as we
made freely available to you the results of research in other fields, such as
radar and jet propulsion."[76] Immediately after the bombs were developed and
dropped, the Americans had changed their attitude and demanded a new postwar arrangement. Those new arrangements, Attlee explained, were agreed to
in the statement of November 16, 1945, "after several discussions by the leaders and their aides." (In other words, the memorandum was not thrust before
the leaders for their blind signatures; they understood what they were signing.)
He did not believe that cooperation was inconsistent with their public declaration on the international control of atomic energy. It was not inconsistent
because they were merely continuing a cooperation that had existed during

the war until a wider system replaced it. Both countries, he concluded, had not deemed it necessary to abandon their joint control of raw materials; thus, they should not abandon all further pooling of information.[77] His response changed nothing. Truman simply ignored the subject. It was left to their respective aides to continue the battle.

What about Truman's response of April 20? Was he as ignorant as he claimed? Even Byrnes had admitted at the CPC meeting of April 15, 1946, that he had been totally unaware of the discussions and of the Memorandum of Intention signed by Anderson and Groves until a short time before the CPC meeting[78] (when Groves introduced it to him?). We know that he did delegate his responsibilities to revise the agreement to Patterson. It is possible that he had not seen the long Memorandum of Intention. In fact, after the April 15 meeting, he had Patterson send him a narrative of the November 16 discussions.[79] Lilienthal reports in his journal that neither Byrnes nor Acheson was aware of the Memorandum of Intention.[80] Byrnes also claimed at the CPC meeting of February 15, 1946, that he had not previously seen the short statement signed by Attlee and Truman.[81] All of this does seem to indicate that Byrnes cared little about the atomic energy side of his foreign policy and did little to keep himself informed. The president, too, claimed that he was not aware of the discussions between Groves, Anderson, and Patterson at the War Department. But Groves did tell Acheson that Truman had instructed him "to go off with one of the British and make specific the broad and general memorandum they had drawn up for future cooperation."[82] Makins would comment that the president's disclaimer "was technically untrue. . . . However, it is quite possible that the President did not fully appreciate full import of the directives he was giving."[83] Did he really believe that he had agreed to a limited form of exchange, or had he naively agreed to a broad exchange, only to be shown, sometime later, the full significance of "full and effective"? Acheson was probably correct when he said that "commitments, on paper and communiqués, have been made and are being made . . . without a knowledge of what the hell it is all about—literally."[84]

Since both Byrnes and Truman argued that they knew nothing about the meeting between Groves and Anderson, one should conclude that the meeting did not have their official blessing. If so, should more credence be given to the agreement drawn up at this unofficially constituted meeting than to the tripartite agreement signed by the two prime ministers and the president? Should not the terms of the relatively more specific short statement weigh more heavily than those of the Memorandum of Intention? The U.S. administrators accepted the terms of the longer memorandum, craftily worded by Groves, to extricate themselves from any binding ties to full collaboration. Moreover, the fact that both Byrnes and the president claimed ignorance of the existence of the Memorandum of Intention would indicate that the phrase "full and effective cooperation" meant just that to Truman, since by his own admission, he was unaware of the qualifying changes introduced by Groves. As Anderson put it, "The tripartite agreement was a simple straightforward document which meant exactly what it said."[85] We have here a president and a secretary of state who turned over to the government bureaucracy the responsibility of formulating foreign policy in an area of vital international and national importance. This trend would continue. American postwar atomic energy policy would be formulated, for the most part, not by the president but by this bureaucracy.

Did atomic raw materials play a role in this current round of debate? Indeed they did. It will be recalled that the memorandum, through Groves's efforts, had proposed the allocation of raw materials on "an actual use basis." It will be recalled also that Byrnes had suggested on February 15 that the subcommittee's report be examined by the respective countries' legal minds. Hence, the entire report was placed in abeyance, including the articles dealing with the allocation of raw materials. At the April 15 meeting, when Lord Halifax introduced the prime minister's ingenious compromise, he also introduced a second proposal addressing the allocation of raw materials. The British demanded an equal allocation of the Congo uranium. Was this an attempt on the part of the British to use raw materials to force collaboration on the United States, now enjoying the fruits of both countries' cooperative endeavors in the Congo? Not really. This proposal was not a knee-jerk response to the Americans' stonewalling tactics; it was carefully thought out.

Almost immediately after the war, the consensus among British officials was that an atomic pile would be established in the United Kingdom.[86] It was also accepted that this British effort would require a share of the Congo uranium. The British did not conceal their plans from the Americans. As early as February 16, 1945, Churchill had informed Roosevelt, in the presence of Hopkins, that the British planned to start up their own project.[87] After Attlee came to power, and in the midst of his communication with Truman to set up a summit, he too had instructed Lord Halifax to inform the CPC, at its October 13 meeting, that the British government proposed to set up a research establishment in the United Kingdom to deal with "all aspects of atomic energy." The proposed establishment included a pile to provide material for research and development.[88] And as late as mid-February 1946, the British informed Groves of their plans, the scope of which, one may add, disturbed him.[89] Despite these research plans, the British still were willing to allow the

Americans full allocation of the Congo uranium until such time as British needs escalated, provided that the Americans were willing to engage in full collaboration.[90]

By February 15, Halifax had already circulated a memo that demanded a more equitable distribution of uranium. The wartime agreement, he explained, had turned over the Congo uranium to the United States for "the production of weapons against the common enemy in the present war." Since the war was over, the uranium aspect of the agreement also had to be reviewed. He proposed that all raw materials received by the CDT since V-J Day and for the period up to December 31, 1946, should be divided on a 50–50 basis between the United States on the one hand and the United Kingdom and Canada on the other.[91] The United States soon countered that all materials received by member nations before March 31, 1946, should be allocated to those governments and that after March 31 the United States should be guaranteed 250 tons of uranium per month to satisfy its plant requirements. The remainder of the materials, it recommended, should be allocated to member governments in accordance with requirements. Halifax rejected the counterproposal on the grounds that it gave a large allocation to the United States and none to the United Kingdom, even though the British were financing 50 percent of the material received by the CDT. Groves was livid. He saw his well-laid plans for the United States' control of raw materials and its continuing monopoly of the "secret" threatened by the intransigence of the British. He argued that the United Kingdom had no current need of raw material for the operation of plants. He then touched upon an issue that would arise again and that more than any other goes a long way toward defining the U.S. reluctance to collaborate with the British. He pointed out that "acceptance of the proposal put forward by the British members would mean a partial shut-down of the US plants that required the maintenance of a long pipeline."[92] The Americans believed that their atomic plants had first claim to all available uranium, and they would continue to hold this view for the next several years.

Two weeks later, Groves again resorted to his heavy artillery in justifying to the U.S. delegates his opposition to the British 50–50 plan. In a memorandum to Acheson, he explained that it was contrary to the principle of need and that if the British were allowed to stockpile raw materials, the effect would be a shutdown of U.S. plants in two years. He added that the United Kingdom should not be encouraged to build a plant that would deprive the United States of part of the limited supply of available raw materials.[93] He continued that in a future war a U.K. plant would be susceptible to enemy bombing attacks and that a 50–50 division would, in effect, punish the United States, which would be using its stock in the present to produce bombs for the defense of peace while the British would be stockpiling theirs for future commercial use.[94]

Meanwhile, the inner circle of the cabinet already had informed Makins of its decision to suspend delivery of Congo uranium to the United States if an allocation decision was not made within a matter of days. It was clear to the cabinet that the United States intended to dominate the world of atomic energy by collaring all available raw material.[95] From Washington, Lord Halifax was even more extreme; he suggested liquidating the CDT if there was no result in five days.[96] On that same day, he told Acheson that his government had been seriously disturbed about the failure to agree on cooperation and that he could not predict what the effect of a second failure on allocation would be but that "it might very possibly be far-reaching."[97] Acheson went back to Patterson, Groves, and others, among whom the matter was debated. John Hancock, a member of the U.S. delegation to the UN Atomic Energy Commission, wrote in reference to the United Kingdom's position that "there are some grounds for argument, but I think both men [Acheson and Patterson] are in the mood to rule Groves out because they feel the British have a good case."[98] Moreover, the U.S. members agreed that they could not risk breaking up the CDT. They needed the support of the British to help them control the sources of raw material.[99] The result was a quick compromise that accepted the basis of a 50–50 distribution of uranium ore. The British threat thus earned them an equal share of Congo uranium. This equitable allocation would result in a huge stockpile in the United Kingdom. This stockpile reentered the collaboration debate in the following year.

By June 6, after the prime minister's letter to Truman, collaboration between both countries was reduced to the joint control of raw materials and nothing more for the next eighteen months or so. This reduced collaboration was due primarily to the efforts of one man: General Groves. There is no doubt that Truman was fully prepared to collaborate with the British. When he gained office, he was led to believe that the British were familiar with all the "know-how" to construct a bomb. In his mind, postwar collaboration would be a continuation of the close atomic association that had existed during the war. Hence, it was not necessary for him to examine all the ramifications of the memorandum he had signed. Byrnes, his secretary of state and foreign policy adviser, due to his own lack of interest, did not raise a red flag. Not until Groves shook them all up did reappraisal set in. Groves was determined to concede nothing to the British. Gordon Arneson reported that on one occasion, just before the U.S. negotiating team left for a negotiating session with the British delegation, Groves instructed them to adopt a tough negotiating stance, "to promise nothing and yield nothing but to do it in a polite manner so as not to offend the British." The session, Arneson added, was conducted along Groves's guidelines.[100] On the other hand, Attlee believed that collaboration was a right that the British had earned through their contributions to the bomb effort. For him, it was a difficult realization that the Americans were unwilling to share information that was as much British as American. Attlee's atomic guru was Anderson, who allowed himself to be manipulated by Groves. Had Anderson not agreed to the inclusion of the qualifying statement in the Memorandum of Intention, doubtlessly the Americans would have created another justification to end collaboration. But he did agree, and it was used. Unfortunately, the British did not quite appreciate how important a bargaining chip raw material really was. The threat to withdraw from the CDT and to withhold Commonwealth ores was effectively used to force equal allocation. It is possible that this threat could have been more forcefully applied earlier in the year to gain a freer exchange of information. By midyear, however, other factors had emerged as seemingly insuperable barriers to collaboration: the passing of the McMahon Act and the efforts to control atomic energy at the international level. Those issues will be discussed in the next chapter.

FIVE

The United States Debates National versus International Control

While the administrations of the United Kingdom and the United States wrangled over the issue of collaboration, the public, media, and elected officials waged another battle. This debate centered on concerns about the future control of atomic energy in a world at peace. The debate grew very heated in the United States, where arguments of patriotism, separation of power, and international diplomacy influenced its outcome. To gain a better understanding of Anglo-American collaboration in 1946, the reader must not lose sight of this unfolding debate because its resolution left a lasting impression upon the atomic energy decisions made by both the United States and the United Kingdom. In the end, the U.S. Congress passed the McMahon Act, which introduced stringent controls over atomic energy developments in the United States. At the same time, the Truman administration attempted to give the United Nations some control over the proliferation of atomic weapons. Anglo-American atomic energy cooperation would be caught in the middle of these two initiatives. The success of national control seemed to kill any hope of collaboration, but the failure of international control breathed new life into the effort.

The American public, and even members of Congress, had no doubt who should control the atomic bomb. As far as they were concerned, it was an American enterprise created by American expertise and finance, so the "secret" of its creation should not be shared with other nations. Public opinion surveys at the end of September 1945 revealed that 70 percent of the citizens and 90 percent of congressmen questioned objected to sharing the "secret" with other nations. Revealingly, more than 80 percent of the same respondents did not believe that the United States would be able to exercise a monopoly of the "secret" for more than five years.[1] The administration was aware of public and congressional sentiments. According to its own confidential opinion poll in mid-October, 70 percent of those polled were still unwilling to share the "secret"; however, 60 percent indicated a willingness to share it if all the bombs were controlled by the United Nations.[2] The administration credited this qualified shift in public opinion to the lobbying efforts of those scientists who had worked on the project.

Scientists were almost united in their opposition to American monopoly of atomic weapons. Their support for international control had started during the war[3] and continued during the postwar period. They did not think that it was possible to exclude the international scientific community from the secret. They saw national control as an attempt to monopolize the so-called "secret" of the atomic bomb. Many of them, including the Federation of Atomic Scientists, set about educating the public in addresses before Congress and at press conferences on the fallacy of an American secret. During the week of November 10, a group of scientists called a press conference in Washington attended by over thirty reporters. The group hammered home to the reporters, among other observations, that the British and Canadian scientists who had worked at Los Alamos knew as much about the bomb as the Americans did.[4] In the interest of their own national security, scientists argued, other nations would be forced to pursue atomic research, the U.S. monopoly would be short-lived, and an arms race would follow. Their solution, for the most part, was international control to monitor the proliferation and use of atomic energy and atomic raw material. Throughout the country, newspaper opinions were divided. The *New York Herald Tribune,* for example, reported on August 8 that "this weapon represents a force too awful for anything less than international control."[5] On the other hand, the *Denver Post* stated on September 10 that "no reliance can be placed on international agreements. . . . The secret of the atomic bomb is safe only in the exclusive keeping of the United States."[6]

On September 21 and 26, Truman polled his cabinet on their views regarding national versus international control and discovered that the split in the country at large was reflected among his cabinet members. Stimson, then retiring as secretary of war, submitted a written opinion four days after the first meeting. He opposed UN control because he could not see the Soviets putting atomic weapons under the control of small UN nations. Instead, he suggested a triumvirate consisting of the United States, the United Kingdom, and the Soviets working together to control atomic energy for industrial and humanitarian purposes.[7] Acting Secretary of the Interior Abe Fortas favored a free interchange of all scientific information with all nations because "it is idle to talk about maintaining a United States monopoly of information about atomic energy. We have no such monopoly, and if we did have it, we couldn't maintain it."[8] Undersecretary of State Dean Acheson agreed. In a radio discussion, he admitted that "the extremely favored position with regard to atomic devices which the US enjoys at present is only temporary. It will not last."[9] Henry Wallace, secretary of commerce, argued that the United States should share its scientific knowledge (not industrial blueprints and engineering know-how) with the Soviets in order to win their trust and create durable international cooperation.[10] James Forrestal, secretary of the navy, disagreed. He described Wallace as "completely, everlastingly and wholeheartedly in favor of giving [the bomb] to the Russians."[11] Clinton Anderson, secretary of agriculture, argued that the "secret" should not be shared with any country, ally or otherwise.[12]

From as far away as the Soviet Union, opinions poured into the White House. On September 30, George Kennan, an American diplomat in Moscow, expressed his views on sharing atomic information with the Soviets in a dispatch to Washington. He explained that it was his profound conviction that "to reveal to the Soviet government any knowledge which might be vital to the defense of the United States, without adequate guarantees for the control of its use in the Soviet Union, would constitute a frivolous neglect of the vital interest of our people."[13] Kennan's distrust for the Soviet leadership and distaste for all that it represented were well known within State Department circles. He was destined to play a major role in influencing American policy toward the Soviet Union. His so-called Long Telegram of February 1946, which clinically dissected the motives and strategies behind Soviet foreign and domestic policies, would provide the justification for the U.S. containment policy.

Across the Atlantic, the British cabinet was equally divided, but its division was not over national versus international control. It was united in its opposition to American national control because it perceived that any such control would deprive Britain of the benefits of cooperation. Rather, the British debate revolved around the issue of international control and the Soviets' role in that new atomic order. The question of Anglo-American control was viewed as impractical. In late August, London instructed its Washington embassy to advise the U.S. president that there could be no Anglo-American monopoly of atomic weapons "since any country with a reasonable number of first-class scientists could solve the problem involved in its manufacture."[14] This was the view of all knowledgeable scientists. Also, it partly explains why some members of the cabinet favored sharing atomic information with the Soviets. They believed that if the Soviets did not already have the bomb they would soon develop it. At a reception at the Soviet embassy, during the September 1945

Conference of Foreign Ministers in London, Molotov, the Soviet foreign minister, implied that they had it. Ernest Bevin, the British foreign minister, reported to Hugh Dalton, chancellor of the Exchequer, who had left the reception earlier, that Molotov, who "had now drunk rather much, even for him," raised a toast in which he said, "Here's to the Atom Bomb." Then he added, "We've got it."[15] Whether it was a drink-induced slip or a deliberately planted seed, it reaped results. Dalton and a few others, whom Dalton did not name, discussed the larger implications of the statement and were "inclined to think that the least risk would be to tell the Russians all we know."[16] They thought it was the only way to break down the growing Soviet suspicion of the West.

Attlee did not agree. His view was that offering the Soviets practical atomic information would not induce them to change their attitude to world problems. The offer would be regarded as a confession of weakness. Rather, he believed that the establishment of better relations should precede the exchange of technical information.[17] Ernest Bevin agreed with him. Further, Bevin suggested that atomic weapons should not be considered in isolation from other weapons; any exchange of information should be considered in the context of all new weapons of war. He proposed that at the upcoming Washington summit with President Truman the two leaders should consider asking the United Nations to devise a system for the full disclosure and exchange of all scientific research on all new weapons to every member of the United Nations.[18]

For the most part, the cabinet agreed with the prime minister's assessment. On November 8, he clarified his views even further. It was virtually impossible for an international organization to control the proliferation of atomic weapons, since no system of international inspection was likely to be workable. Therefore, no attempt should be made to restrict the development of atomic energy by any country. The peace of the world should be preserved by a United Nations composed of all peace-loving nations, willing to act collectively against aggressor nations—with atomic weapons if necessary.[19] In Attlee's new world, nations would renounce war, then set about arming themselves with atomic weapons, in the event that the United Nations called upon them to destroy an aggressor who had used atomic weapons on another. The Chiefs of Staff concurred. The best method of defense against atomic weapons was the deterrent effect that the possession of the means of retaliation would have on a potential aggressor. They recommended that the British "should press ahead in the field of research and that it is essential that British production of atomic weapons should start as soon as possible."[20] The British then were not interested in any form of international control that would deprive them of the security provided by the possession of atomic weapons. Atomic deterrence would evolve to become the cornerstone of Britain's atomic policy.

While the U.S. and U.K. leaders were formulating their future atomic policy, the members of the U.S. Congress hastily began addressing their own atomic policy. They agreed on the need to establish some sort of domestic control over the U.S. atomic energy program. By September 23, senators were already lining up to introduce atomic resolutions. Senator Elbert D. Thomas of Utah had one. So too did Senator Arthur Vandenberg, a Michigan Republican with isolationist sentiments. He supported a strong but independent U.S. foreign policy backed up by a powerful military. The United States' control of atomic weapons provided it with the means to engage in such a policy. He conceded that the United States would not be able to maintain the "secret" for more than five years but rationalized that during that five-year window the United States should use its monopoly to pressure the world into accepting international control.[21] The congressional momentum gained urgency when Wallace's position at the September 21 cabinet meeting was leaked to the press (probably by Forrestal) and misrepresented to the public. The media portrayed him as willing to give the technical secrets of manufacturing the atomic bomb to the Soviets. Truman denied the allegations but saw the need to get his own recommendations on the floor before any senator introduced "harmful resolutions."[22] This was followed by his address to Congress on October 3, in which he proposed the creation of a U.S. Atomic Energy Commission.

The May-Johnson Bill was already in the works before the president's address. In July 1945, Kenneth Royall and William Marbury, two War Department attorneys, had written a draft of an atomic control bill. Among other things, the draft called for the creation of a part-time commission of four military and five civilian members. Each of the two groups could exercise a veto, and members would serve indefinitely. The commission would control all aspects of atomic work in the United States. Groves and Patterson supported the draft for good reasons. The draft placed the military in a position to continue its control of atomic energy with the exercise of its veto powers. On October 4, the day after the president's address, Congressman Andrew May and Senator Edwin Johnson introduced the draft to the House as the May-Johnson Bill.

The bill quickly ran into opposition. Scientists opposed the implicitly dominant role proposed for the military. For example, the newly formed Federation of Atomic Scientists, whose members had worked under Groves, foresaw a continuation of tight military security. The scientists were concerned that the free exchange of information among scientists would be rigidly controlled, thereby adversely affecting further development in the field.[23] In fact, the bill did threaten imprisonment or fines for those who breached its security provisions. Despite the opposition of the rank and file, Oppenheimer, Fermi, Lawrence, and Compton supported the bill. Oppenheimer rationalized that "it was safer to trust the commission and the administrator to use their powers wisely than to risk the chance that Congress, subjected to multiple pressure during a long debate, might produce something very much worse."[24] That argument did not sway too many scientists.

Initially, the president tacitly supported the May-Johnson Bill. However, political reality quickly set in. The president soon became aware that under the provisions of the bill his power could decline the further removed he was from the bomb. He reassessed the bill and concluded that in its present form it contained "a number of undesirable features" that required substantial amendments.[25] In particular, he took exception to the section governing military representation on the commission. Patterson, however, refused to compromise. Meanwhile, Senator Brien McMahon, a freshman Democrat from Connecticut and a strong adherent of civilian control, submitted a rival bill to the Senate. The new bill, the McMahon Bill, proposed a nine-member, full-time civilian Atomic Energy Commission whose members would be nominated by the president. Groves bluntly refused to cooperate with the senatorial committee drafting the bill. In the face of this intransigence by both Patterson and Groves and the loss of momentum for the May-Johnson Bill, Truman shifted his support to the McMahon Bill. Throughout January 1946, the military and its supporters mounted a national counterattack against the new bill. On February 1, after a two-hour session with Senator McMahon, Truman gave him a strong letter of support for the bill[26] and then released it to the public.[27]

But Groves would not be beaten. The day after the press release, Drew Pearson, an American radio commentator, leaked the existence of a spy ring in Canada. On February 16, three days after Groves had taken his Anglo-American collaboration fears to Byrnes, the Canadian government, in the climate of what had become a major scandal, prematurely arrested twenty-two suspected spies. The existence of the spy ring was not news to the Canadian government. In September 1945, the government had informed Truman that it had uncovered a Soviet atomic spy ring in Canada. Herken related that the president was not alarmed since Soviet spying was common knowledge to U.S. officials. Moreover, Groves did not believe the Soviets had the material means to produce a bomb.[28] Hence, the spies had not been arrested in order to flush out the ring. Almost simultaneously with the Canadian arrest, Frank McNaughton, a Washington columnist, revealed that a second spy network was operating in the United States. American and Canadian authorities suspected that Groves was behind the leaks, and some years later McNaughton corroborated their suspicion.[29]

The spy scare worked to Groves's expectation. Senators concluded that civilians could not be trusted to protect the "secret." Senator Vandenberg, a Groves ally, introduced an amendment that created a military liaison board to act as a "watchdog" over the Atomic Energy Commission (AEC). Any doubts the congressmen might have had that they had done the right thing were removed a few months later. The British announced the arrest of Alan Nunn May, a lecturer at Kings' College in London, who had worked on the atomic project during the war, mainly in Canada and England. He was charged with passing atomic information to the Soviets.[30]

When the bill became law in the Atomic Energy Act of July 1946, it specified, among other provisions, the creation of a five-member civilian commission and a separate military liaison group, to be appointed by the secretary of defense and consisting of two representatives from each of the three armed services. Groves was appointed its first chairman. The act also created a special congressional committee, the Joint Congressional Committee on Atomic Energy (JCAE), composed of nine senators and nine representatives, to act as a watchdog over both military and civilian groups. The act prohibited the United States from sharing industrial atomic information until both Houses of Congress were satisfied that effective and enforceable international safeguards were in place to prevent its use for destructive purposes. Furthermore, atomic weapons information could be disseminated by the AEC only if such information assured the "common defense and security" or did not adversely affect the common defense and security.

The quick passage of the act by the Senate caught the British by surprise. They were still in the midst of formulating a policy to address changes anticipated by congressional adoption of the bill. The British were aware that the McMahon Act would have an adverse effect on collaboration. Attlee had complained to Averell Harriman, the U.S. ambassador to London, that passage of the bill would "prohibit the disclosure or sharing of atomic secrets with any foreign power, including the British." He believed that the bill should recognize the mutual cooperation that led to the development of the bomb.[31] Hence, the British were aware of the restrictive nature of the bill. By May 10, their legal advisers had already submitted to the cabinet their conclusion that the bill would end not only the exchange of atomic information but also the exchange of fissionable material that was physically in the United States.[32]

Attlee and Anderson discussed ways to minimize the restrictive effects of the bill. Initially, Anderson suggested that the prime minister persuade Truman to honor the terms of the Quebec Agreement before the House of Representatives passed it[33] or intervene either to prevent passage of the Bill or to secure the insertion of a special clause providing for collaboration with the

United Kingdom and Canada.[34] Makins did not believe that any such approach would persuade Truman to attempt to affect the intent of the bill.[35] Neither did Chadwick, who suggested instead that they should focus on persuading Truman to honor the memorandum of November 16, 1945.[36] In the end, the prime minister refrained from giving the Americans any lead as to how the problem of collaboration should be resolved, since only the president could make that judgment. Rather, he sent off his much-delayed long telegram of June 7, which outlined the history of Anglo-American collaboration, the intent of the November memorandum, and the British displeasure. Of course, the letter did not have its intended effect. Truman simply ignored it, and the bill made its way through Congress.

Why did Truman not take the initiative and intervene on behalf of the British? Up until the passage of the McMahon Act, the administration never did mention to Congress the existence of the Quebec Agreement, the Hyde Park Aide-Memoire, or the November 16 memorandum. The congressmen who voted on the McMahon Act knew that the British and Canadians had made some contributions to the bomb project. But they were totally unaware of the extent of that contribution and did not know that the contributions had been formalized in international agreements predicated on quid pro quo arrangements. Six years after the passage of the act, Senator McMahon would concede that the British did contribute "heavily to our own wartime atomic project." However, he continued, "due to a series of unfortunate circumstances, the nature of the agreements which made this contribution possible was not disclosed to me and to my colleagues . . . at the time we framed the law."[37] Undersecretary of State Dean Acheson did not learn of the existence of the Quebec Agreement until February 1946, and when he did, what disturbed him the most was the potential effect of its existence upon relations with Congress. He mused that the administration had been engaged for some time with discussions with Congress "without mentioning the existence of so relevant and important a matter as this agreement." However, he explained, the cover-up was not a deliberate attempt to mislead Congress; "the somewhat incredible truth was that few people within the administration knew of its existence, due to the many changes in government."[38]

The evidence does not support Acheson's argument that a cover-up was not intended. We have already seen that a conscious decision had been made in London and Washington to keep the terms of the agreement secret for various reasons already discussed. To the credit of the English Parliament, it was able to sniff out the possible existence of the Quebec Agreement. But Attlee denied its existence. Even after relations became bumpy, by February, the British, due to the overriding factors against disclosure, stood firmly against a public disclosure. The cover-up served everyone's purposes. Acheson did add an additional reason for secrecy. He speculated that Truman did not tell Congress of the U.S. obligation to the United Kingdom because he had no intentions of fulfilling this obligation. Hence, he stood back and allowed Congress to take him off the hook by legislating against the sharing of information.[39] One may also argue that Truman did not intercede because he recognized the futility and the risk of any such intervention. He was probably unwilling to jeopardize the passage of the McMahon Bill. By the end of January, Truman's reassessment had led him to recognize the advantages (from the executive standpoint) of weakening the influence of the military and eliminating the veto provided for by the May-Johnson Bill. He realized that any injection of the British role in the development of the bomb and any attempt to seek some sort of accommodation for them might have tilted the congressional balance in favor of the May-Johnson Bill. Thus, it was imperative to keep the British partnership a secret.

Moreover, one has to keep in mind that the British had never enjoyed a great measure of popularity among the American public at large. The favor with which they were viewed during the war was quickly dashed after Attlee and the Labor Party replaced the immensely popular Churchill in 1945. The American public viewed Churchill's defeat not only as a betrayal of their wartime hero but also as a betrayal of and an attack on the free enterprise system. The British were aware of American sentiments. Hugh Dalton confessed that at the end of 1945 relations between the United Kingdom and United States were none too good: "I have the sensation that the Democratic Party in the USA is reverting to what it used to be before Roosevelt's time, with a strong Irish-American flavour, and not much sympathetic understanding towards us."[40] Kenneth Harris described this anti-British sentiment by explaining that when John Maynard Keynes went to Washington, in September 1945, to negotiate a postwar loan for Britain, it soon became quite obvious to him that the American public was not interested in the British sacrifices that had been made during the war. "There was a widespread feeling that American free enterprise was being asked to finance British socialism."[41] When Attlee visited in November, he too was forced to come to grips with the anti-British feelings. On November 13, he addressed a joint session of Congress in which he tried to remove some "misimpressions" about socialism by describing the nature and philosophy of his Labour Party. Keynes himself was told by the American negotiators that they were preparing stiff terms for the loan only because Congress would have to authorize it and that Congress would be wary of lending money to a foreign country governed by a political party committed to the abolition of private enterprise.[42]

Attlee did not blame the president for his failure to stand up for the British. He believed that Truman had to pay a price to keep the ultimate control of atomic energy out of the hands of the generals and that the price was to accept an act that forbade cooperation with foreign powers. Attlee declared that "it wasn't Truman's fault. Congress was to blame."[43] Was Congress to blame? Had they known about the secret agreements, would they have accepted an amendment to include an exemption for the British and the Canadians? We do not know. We do know that Groves and Senator Vandenberg worked together, and, knowing what we know of Groves, it is not beyond the realm of speculation that Groves might have used his influence to defeat any such attempt at accommodation. In response to a comment by Lilienthal that the British were upset by the U.S. denial of information to them, General Dwight Eisenhower revealed that the British blamed Groves. They believed that he was responsible for "going around behind their backs and having that provision against exchange of information put into the McMahon Act, even pointing to the line which singles out 'industrial uses,' which shows on its face that whoever got that line inserted had the background which only Groves and two or three others had."[44] By July 1946, collaboration in the area of atomic energy information was dead. Fortunately for the British, they were able to gain an allocation agreement that provided them with Congo uranium. This uranium would be used to start up the British program.

The international control of atomic energy was the second issue that played a decisive role in the collaboration debate of 1946. When Attlee and Truman met in Washington in November 1945, both men gave the international control issue serious consideration. The result was a public declaration to assure the world that they were considering international action to control the use of atomic energy. However, for any such control movement to succeed, Soviet support had to be won. This was no foregone achievement, since the Soviets had already interpreted the Washington Declaration as another attempt on the part of the British and Americans to gang up on them. The United States and the United Kingdom had a long-standing atomic partnership from which the Soviet Union was excluded. Naturally, the Soviets were suspicious. For the U.S. part, Truman, some members of his cabinet and Congress distrusted the Soviets' resolve to agree or live up to any agreement. Moreover, the Soviets had left no doubt that their and the Americans' interpretations of "democracy" and "free elections" in the liberated areas of Eastern Europe differed on ideological grounds. U.S. officials were aware that Soviet-controlled groups in Bulgaria and Romania were taking steps to determine the outcome of "democratic" elections. They detected Soviet intrusion, though less aggressive, in Finland, Hungary, Austria, Iran, Czechoslovakia, and Greece. The wartime alliance of convenience was crumbling quickly around the United States, and although the United States possessed the atomic bomb, it seemed to be helpless to determine where the pieces fell.

All the same, on December 15, Byrnes landed in Moscow with an American proposal for mutual Soviet-American cooperation on atomic energy under the auspices of the United Nations. Terms of the proposal included communication among atomic scientists in the areas of basic scientific information; exchanges of technological, raw material, and engineering information; and safeguards against the use of atomic weapons. At this Council of Foreign Ministers meeting, Byrnes and Soviet Foreign Minister Molotov, joined some days later by Stalin, hammered out a compromise agreement. The Soviets agreed to support the formation of a UN atomic energy commission in exchange for Byrnes's concession that the Security Council, where the Soviets could exercise its veto, would control the agency. The Soviets then conceded to Byrnes's demand that the terms of the proposal be implemented in stages, after arrangements for inspection and safeguards were worked out.

Truman quickly appointed a five-man committee under the chairmanship of Acheson, and including Bush, Conant, Groves, and former Assistant Secretary of War John J. McCloy, to devise an international atomic energy plan for submission, ultimately, to the United Nations. A board of consultants under the chairmanship of Lilienthal advised the committee. Acheson and his committee took their assignment seriously. By March they had drawn up some guidelines to be presented to the United Nations. The so-called Acheson-Lilienthal report included the following recommendations: (1) that nuclear raw materials and all industrial plants producing fissionable materials be internationally controlled; (2) that atomic energy be divided into "safe" and "dangerous" categories; and (3) that all nations be held accountable for all supplies and location of fissionable materials.[45] The report also raised the fear that future enforcement procedures to control atomic proliferation might prove inadequate.

Meanwhile, Truman and Byrnes began their search for a man to represent the U.S. interest at the United Nations. They sought someone of international and national prestige who would command respect at home and abroad.[46] He had to be a man of conservative outlook so as to reassure Congress and the public that he would not give away the "secret." They nominated Byrnes's close and personal friend Bernard Baruch to be the U.S. representative to the newly formed UN Atomic Energy Commission. Lilienthal commented that he was quite sick when he read the news of Baruch's appointment. The choice "seemed to me fantastic—his age, his unwillingness to work, his terrifying vanity." When all these things were taken together, Lilienthal concluded that

Baruch could not win over the Soviets.[47] In his memoirs, even Truman described Baruch in the following unflattering terms, as a man who was "usually referred to as an advisor to Presidents. . . . His concern, in my opinion, was really whether he would receive public recognition. He had always seen to it that his suggestions and recommendations, not always requested by the President, would be given publicity. . . . Baruch is the only man to my knowledge who has built a reputation on a self-assumed unofficial status as 'advisor.'"[48] Yet Truman appointed him. Baruch himself admitted that in scientific circles there was widespread disappointment and dissatisfaction over his appointment because of his nonexistent scientific background. Bush, for example, his friend, told him "point blank that, in his opinion, I was the most unqualified man in the country for the task."[49]

On June 14, 1946, Baruch submitted his plan to the United Nations. It was similar to that drawn up by the Acheson-Lilienthal committee but with stringent enforcement measures added. It was these provisions that would contribute to the failure of the plan. They included the following: any nation that violated the terms of the UN agreement "would be liable to immediate, swift and sure punishment." The use of atomic force was not ruled out but was implied. The plan also stipulated that nations would be unable to use their veto to "protect those who violate their solemn agreements not to develop or use atomic energy for destructive purposes." This article was included to prevent the Soviets from vetoing enforcement at a time when the United States commanded a clear majority in all UN bodies. Further, only after the United Nations had put an effective control system into place would the United States cease the production of bombs and dispose of its existing stock.[50] In his memoirs, Baruch summarized the U.S. proposals in the following terms: "It was a proposal unprecedented in its generosity. The United States offered to surrender a weapon it alone possessed, and to share its secrets, on only one condition—that every nation be guarded against its destructive use."[51] James Byrnes agreed with Baruch's assessment; he stated: "History will not disclose action by any government comparable to this generous offer. . . . We could have held this weapon and used it as a threat to force concessions from other governments. Instead, the three governments hastened to offer the secret of their power to an international organization in order to have it used for peaceful rather than destructive purposes."[52]

Unfortunately, others did not quite see the offer in that light. Lilienthal described the proposal as one containing "some absurd stuff about sanctions and penalties."[53] Oppenheimer "was deeply troubled by the sanctions talk and by the implications of it."[54] The Soviets too were not impressed by the Americans' magnanimous offer. What they saw was a nation that possessed the technology to produce atomic weapons taking steps to prevent them from developing that same technology and a nation that already had bombs in its arsenal while they had none. They saw a nation that exercised a great deal of influence in a United Nations located in its own backyard. In particular, they saw a nation that did not look kindly on the Soviet system of government and would not hesitate to destroy it. The Soviets had no intention of shutting down their atomic program until they had produced their own arsenal of atomic weapons. In fact, during the Potsdam Conference of July 1945, the day after Truman had implied to him that the United States had produced an atomic bomb, Stalin ordered his chief atomic scientist, Igor Kurchatov, to accelerate the pace of the Soviet bomb project.[55] One year later, his atomic scientists were working feverishly to produce a bomb. Stalin saw nothing in the Baruch Plan to induce him to curtail their research.

The British were not overly excited either. In his oral memoir, Lincoln Gordon, a consultant to Baruch, reported that some of the younger members of the U.K. delegation had confessed to him that when they first read Baruch's proposal they were skeptical. "They thought this was some sort of ploy, and it took them some time to take it seriously.[56] Apparently, Sir Alexander Cadogan, head of the U.K. delegation, never took it seriously. He apologized that he could not play a more active role in the debates because "so much of the debate is what I heard in Geneva for six years [as a member of the U.K. arms control team before World War II], and I am positively pickled with boredom."[57] The British took a realistic approach to atomic energy control. They knew that the Soviets would not agree to the physical inspection of their atomic energy plants and that even if they did agree it was virtually impossible to keep tabs on Soviet plants scattered throughout Eastern Europe.[58] Moreover, they conceded that the Soviets could not agree to the U.S. government's retaining its existing stock of atomic weapons unless the Soviets had equal access to atomic weapons.[59] By July, Bevin was reminding Cadogan that the Soviets placed too high a value on their right to exercise the veto to agree to any veto proposal. Bevin stated that he too opposed the idea of abolishing the veto. His argument was that the United Kingdom might need to use it in the future to protect her interest against the vocal smaller nations of the United Nations.[60] Parliament too did not react very well to the Baruch Plan. The left wing of the Labour Party viewed it as serving the interest of the United States. In a House of Commons debate on August 2, Captain Blackburn surmised that "the effect of the American proposal as they now are is almost to disarm Russia, while America herself remain armed; . . . one could hardly expect the Russians to accept at the outset proposals of that kind."[61]

Despite these serious misgivings, the British officials did not openly criticize the plan or fully exert themselves on its behalf. In reference to the British, Lilienthal commented that "the British are against our plan, though very cagey. They see atomic energy as a source of power, a way of pulling themselves out of their economic troubles."[62] Moreover, the British had already decided that it was in their security interest to own atomic weapons. Both they and the Soviets viewed an American monopoly of atomic weapons with suspicion.[63] But, like a loyal ally, the British tagged along behind the Americans until the last moment, when Cadogan made a feeble and futile attempt to oppose them. On December 30, 1946, just before the UN committee voted on the U.S. plan, he called Baruch aside to announce that the United Kingdom could not support it. Baruch later wrote that "I could feel the color leave my face. I told him in the plainest language, and in the presence of others, that if Britain walked out on us, I would denounce her in terms that would make 'perfidious Albion' sound like words of praise."[64] Britain did vote with the United States. But the plan was destined for the rubbish heap of history, and this was clear from its inception.

Within days of Baruch's introduction of the plan to the United Nations in June, Andrei Gromyko, the Soviet representative, countered with a plan of his own. It called for (1) the absolute prohibition on both the production and use of atomic weapons, (2) the destruction of all stocks of such weapons within three months of the convention coming into force, (3) the enforcement of the terms of the agreement by all states within or without the United Nations, and (4) follow-up action to monitor compliance. International control thus died an early death because neither side was willing to budge. The United States was not willing to destroy its atomic arsenal until it was satisfied that international control would work, and the Russians were not willing to consider international control until the United States agreed to destroy its weapons. On December 31, the Baruch Plan came up for a vote before the Security Council. Ten nations voted to support the plan, but the Soviet Union and its satellite Poland abstained. Without Soviet endorsement the plan was dead. Neither side was willing to be the first to walk away from the negotiations, so the debate dragged on for another year of futile modifications and additional debate.

By mid-1947, it was clear to all concerned that international control would not be realized. An intelligence report in late 1946 had already concluded that the Soviets intended to press for international disarmament and the outlawing of atomic weapons while, beyond the reach of UN inspectors, they planned to continue their own atomic development.[65] This breakdown contributed to the U.S. reappraisal of its relationship with the British. In November 1946, Fred Searls, who had seen the futility of international control and had resigned from Baruch's delegation, suggested to Byrnes a plan to prevent the Soviets from gaining access to raw materials. The scheme called for the formation of a group comprising as its core the United States, Canada, and United Kingdom that would control atomic energy through its monopoly of raw materials.[66] Searls was not the only one who saw the need to devise alternate plans to deal with the inevitable breakdown of international control. Shortly before he left his post as undersecretary of state, Acheson too began looking beyond the failure of international control. He asked George Kennan, now back from Moscow and head of the Policy Planning Staff, to draw up contingency plans. On August 21, 1947, Kennan submitted a report that dismissed the fourteen months of talks in the United Nations as fruitless. He advised that American policy on atomic energy should not be pursued through the United Nations. Rather, the United States should face the grim realities and force a return to close relations with Britain and Canada.[67] The grim reality was that atomic weapons would not be outlawed, so a policy had to be devised to deal with the Soviet Union's eventual possession of atomic bombs.

The State Department saw the need to reevaluate its international atomic policy. In a letter to the U.S. ambassador to Belgium on June 27, 1947, Acheson outlined the State Department's fears. The Soviet's line was to filibuster in the United Nations while at the same time (1) breaking down existing U.S.-U.K. raw material procurement arrangements (this was a direct reference to the Belgian communists raising questions in parliament about the U.S.-U.K. procurement arrangement in the Congo); (2) infiltrating atomic research and control programs in other countries (it was believed that the Soviets had infiltrated the French atomic program); (3) hastening the development of their own program; and (4) extending the areas of their political domination in foreign countries (they were in the process of establishing satellite states in Eastern Europe).[68] By the middle of the year, State Department officials had convinced themselves of the need for some sort of atomic rapprochement with the British. Soviet behavior warranted cultivating the friendship of the British by accommodating them in an atomic alliance. Of course, before any atomic relationship could be entered into, the Atomic Energy Commission and the JCAE, the congressional watchdog, would have to be convinced of its security advantages. The means to win these committees over was near at hand. There was a dire shortage of raw materials to feed the U.S. atomic energy plants.

SIX

The United States Initiates
the Modus Vivendi

The year 1947 opened with a British initiative to revive atomic cooperation. The president and the prime minister, after eighteen months of stepping around the subject, once again reintroduced the words "atomic energy" in their official correspondence. British officials quickly seized the moment and attempted to steer the Americans to the negotiating table. Their initiative failed, leaving them shattered with disappointment. The British, for the most part, resigned themselves to the futility of atomic collaboration. It became increasingly clear to them that any British program might have to be developed primarily on the basis of British know-how. The prime minister, with the best advice of those consulted, made the momentous decision to proceed with an independent British atomic program. However, out of the gloom of noncollaboration shone a new beam of hope directed by members of the State Department and the U.S. Atomic Energy Commission (AEC). These officials came to recognize the security benefits of improving relations with the British, since the British, with their huge stockpile of uranium, seemingly held the key to the Americans' uranium shortages. The State Department felt that atomic ties had to be renewed before the Soviets, growing ever more adventurous in Europe, ended the American atomic monopoly. The Americans offered the British a carrot in the form of renewed atomic energy cooperation, an offer that they readily accepted. Meanwhile, the State Department and the AEC set about winning over the new congressional "watchdogs" to their rapprochement policy. The State Department and the AEC were resoundingly successful in persuading the Joint Congressional Committee for Atomic Energy (JCAE) to give its blessing to Anglo-American negotiations and, at the negotiations, in persuading the British to turn over their uranium reserves and the entire Congo output to the Americans. This agreement, which included other provisions, was called the Modus Vivendi. This chapter will trace the details of the materialization of this new American policy and examine the domestic and international issues that led the British to accept it.

In January 1947, the British made a secret decision to proceed with the development of an atomic bomb. A new ad hoc committee, GEN 163, which replaced GEN 75, made this decision. The committee consisted of Attlee, Bevin, Herbert Morrison (lord president of the Council), Lord Addison (secretary of state for the Dominions), A.V. Alexander (minister of defense) and John Wilmot (minister of supply). The prime minister also sought advice from Lord Portal, Anderson's Advisory Committee, and the Chiefs of Staff. They all supported the decision to go ahead. The full cabinet was not privy to the factors that went into this decision. At a time when Truman consulted his cabinet and encouraged it to express its views on atomic energy matters, Attlee did no such thing. In fact, Stafford Cripps and Hugh Dalton, two of the insiders, who had earlier opposed a bomb project on economic grounds, were not invited to the January meeting. During the six years of the Labour government, 1945 through 1951, atomic energy appeared on the cabinet agenda fewer then ten times. Half of these appearances were during the administration's first six months, and two were on the eve of Attlee's 1950 summit with Truman,[1] when the president implied that he was considering the use of the atomic bomb in Korea.

Likewise, input from Parliament was never solicited. From time to time, Parliament was merely informed of selected highlights about the program. For example, on October 29, 1945, Cripps notified its members that the government was setting up the Atomic Energy Research Establishment;[2] three months later, the prime minister announced that the government planned to begin the production of fissionable material for research purposes. He placed the overall project under the Ministry of Supply, with Lord Portal, wartime chief of the Air Staff, as its administrator with the title "controller of production of atomic energy." He announced John Cockcroft as his choice to direct the research establishment at Harwell.[3] But when he made the all-important decision one year later, the Commons was not informed. The House found out in May 1948 when the minister of defense, in response to a question, revealed that "research and development continue to receive the highest priority in the defense field, and all types of modern weapons, including atomic weapons, are being developed."[4]

The breakdown of collaboration did play a role in influencing the decision to proceed with the project. Since the Americans were unwilling to assist them, it became imperative to the British that they provide for their own security. Attlee stated some time later that "if we had decided not to have it, we would have put ourselves entirely in the hands of the Americans. That would have been a risk a British government should not take. . . . At that time nobody could be sure that the Americans would not revert to isolationism. . . . For a power of our size and with our responsibilities to turn its back on the Bomb did not make sense."[5] The Britain that Attlee referred to was in fact a power of size and international responsibilities in 1946. She was looked upon to assume a leadership role in Europe at a time when no one was certain what America's future European role would be. In retrospect, one may agree that postwar Britain was a nation in decline, but in 1946 it did not experience any economic difficulties that other European nations were not already experiencing. They were all going through the postwar trauma of reconstruction. Unlike the others, however, Britain still enjoyed all the trappings of a great power. It was the accepted leader of a huge family of Commonwealth nations, and it still ruled an expansive colonial empire. One can excuse the British government if, as Roger Ruston stated, it "still deluded itself into believing that it was a great power."[6] In the minds of its leaders and the leaders of the world outside the United States, it was still a great power that, with its American and Soviet allies, had just won a great war.

To maintain its status and be counted among the great powers of the world, it had to possess that new symbol of power: the atomic bomb. As Bevin so succinctly put it in October 1946: "We have got to have this thing over here whatever the cost. . . . We've got to have a bloody Union Jack flying on top of it."[7] The fact that international control was going nowhere only hardened the British resolve. It was clear to the Chiefs of Staff that other nations would ultimately develop the bomb; therefore, the United Kingdom "should not be without this weapon if others possessed it."[8] The way things stood, the Americans exercised sole control over the technical know-how of its manufacture and were pushing an international plan that would tighten that monopoly. The British were not willing to acquiesce in an American monopoly of atomic weapons. The British government was forced to consider the effect on Britain of the breakdown, or, conversely, the success, of international control. It had to address the future plight of a Britain, a great power, without atomic weapons. The U.S. refusal to collaborate had added a sense of urgency to the atomic bomb decision.

The January decision was difficult in more ways than one. Britain's economic and financial structures were on the verge of collapse due to high unemployment, fuel shortages, a freezing winter, and the convertibility of sterling in August 1947. The year would turn out to be, as Dalton called it, his "annus horrendus."[9] Moreover, when the government made its decision to develop an expensive atomic program, it knew that funding would be needed for the welfare state it was building. Money was needed for its National Health Service, housing program, National Insurance, education reforms, and other social programs. Attlee could not justify an expensive atomic program in a period of strict austerity, which included food and clothes rationing. Hence, he had to go the way of secrecy. If he had chosen to develop the weapon by means of the ordinary agencies in the Ministry of Supply and the service departments, it would have been difficult to conceal its development. But if "special arrangements conducive to the utmost secrecy" were taken, the project could be "camouflaged" as "Basic High Explosive Research."[10] It was camouflaged. The government recruited Dr. William G. Penney into the Ministry of Supply and placed the bomb project under his control. He worked in the utmost secrecy, directly under Lord Portal. Dr. Penney had worked on the Manhattan Project, and the Americans had regarded him as the most knowledgeable of the British scientists working on the bomb.

Attlee kept the legislative branch of his government totally uninformed, at a time when Truman, albeit reluctantly, chose to bring Congress in as a partner in the decision-making process. It is quite possible that, in addition to financial considerations, the presence of the many left-wingers in the Labour Party played a role in this secrecy. These left-wingers frowned upon the close alliance with the United States. Some favored closer ties with the Soviets, while others preferred an independent foreign policy. The leftists did not disturb the party too much, since the party's National Executive Committee easily snuffed them out. Some extreme leftists were actually expelled from the party for opposing the government and promoting communist ideology. All the same, Attlee had to be conscious of their presence. They would have protested if they had known of the United Kingdom's treatment by the United States, Attlee's attempt for closer collaboration, and the financial demands of the bomb project in a climate of austerity and social innovations.

The breakdown of atomic cooperation did not adversely affect Anglo-American relations in the fight against the Soviet Union and the spread of communism. By early 1947, Moscow was seen as behind the Azerbaijan revolt in northern Iran, as threatening Turkey, and as exploiting the leftist rebellion in Greece. These fears arose at a time when Britain's financial problems kept it from responding adequately. On February 11, Bevin informed the U.S. State Department that Britain was about to suspend military and economic aid to Greece and Turkey. The Americans responded a month later with the Truman Doctrine, whereby the United States took over the burden of military aid to Greece and Turkey. In Germany, both nations found common ground again by making the historic decision to merge their occupation zones into the so-called Bizonia, despite the futile protest of the Soviets. Cooperation continued in the economic reconstruction of Europe. The Americans saw in the econcmic stagnation of Europe an opportunity for communist exploitation of th:s hardship. They offered the Marshall Plan in June 1947, which the British speedily embraced and promoted, for the economic, and ultimately political and military, development of Europe. Hence, in the midst of atomic animosity and bitterness, cooperation reigned against a Soviet enemy that was rapidly expanding its empire into areas once coveted by a Germany that was, just as rapidly, shedding its stigma of "enemy."

During the course of 1946 and 1947, no such cooperation existed in the atomic field. By the end of 1946, Truman's refusal to answer Attlee's telegram of June had added an "element of discourtesy" to the controversy. The cabinet believed that the refusal should be addressed, if for no other reason than to show that "the conception of full and effective cooperation is still very much alive in our minds."[11] Roger Makins was against any response until the members of the U.S. AEC were appointed and had an opportunity to study the issues. He anticipated no results from repeating the same old arguments to the State Department. "It is a mistake to flog a dead horse, especially when a new one is about to emerge from the stable."[12] All the same, the dead horse was flogged as Bevin "jumped the gun," in late 1946, by raising the issue with Byrnes in New York. Byrnes contributed nothing to fuel British optimism, but the contact initiated a new round of talks. The Washington embassy proposed to London a battle plan that called for the prime minister to send a short telegram to the president referring to his letter of June 6th. The second step was to have the Washington staff follow up the letter with visits to the State Department, members of the Joint Chiefs of Staff, and the AEC.

The plan was duly implemented. In mid-December 1946, Attlee wrote Truman that he had not pressed him earlier for a reply to his June letter because he was mindful of his difficulties in Congress. But since the act had been passed and international control was not imminent, he felt bound to enquire how the president intended to effect their November understanding.[13] The ice was broken. Truman responded that he would consider the matter with his advisers and the AEC and get back to him in the near future.[14]

The following month, January 1947, stage two was put into effect: make the rounds of Washington and win the support of sympathetic officials. General Dwight Eisenhower, a man with British sympathies, was now chairman of the Joint Chiefs of Staff; General George Marshall, another Anglophile, was about to be appointed secretary of state; Dean Acheson was still undersecretary of state; and the AEC had just taken over the American atomic program. The British embassy wasted no time. Makins, who was about to return to London to take up an undersecretary post in the Foreign Office, held informal discussions with Acheson. Acheson offered a sympathetic ear and the advice that the British try to persuade the U.S. military men to support the development of atomic plants in the United Kingdom.[15] Field Marshall Henry Wilson, head of the British joint staff mission, then sought out Eisenhower. Eisenhower in turn suggested that the State Department and not he present the British case to the Joint Chiefs so that he would be better able to support the British claims.[16]

Meanwhile, on January 23, Makins held an informal meeting with Dr. Carroll Wilson, general manager of the AEC, followed by a meeting with the full AEC six days later. David Lilienthal was Truman's choice to be the first chairman of the AEC. The former director of the Tennessee Valley Authority still had to survive a contentious senatorial hearing where detractors would accuse him of mismanaging the TVA and of harboring communist sentiments. The four other commissioners were Robert Bacher, a physicist from Cornell University who had worked at Los Alamos; Lewis Strauss, a wealthy Wall Street financier with close ties to the military; Sumner Pike, a businessman; and William Waymack, editor in chief of the *Des Moines Register*. At that meeting the AEC defined to Makins the line of responsibility between itself and the State Department. The AEC held itself accountable for the raw material operation of the Combined Development Trust, but it regarded the Quebec Agreement and the Combined Policy Committee (CPC) as falling under the jurisdiction of the State Department. It suggested that the foreign affairs departments of both governments work out the collaboration problems.[17] Perhaps the most important point that emerged from the meeting was the commissioners' evident fear that the secret arrangements would leak out. Under the terms of the McMahon Act, they were under a legal obligation to keep the JCAE fully informed on the raw materials agreement. However, they believed that the State Department also had a responsibility to disclose the other secret agreements to Congress.

The British had gone full circle, but they had now a better understanding of the decision-making process and their chances of success. Their optimism grew. They approached Secretary of State George Marshall, and he promptly turned the matter over to Secretary of the Navy Forrestal and Secretary of War Patterson. He instructed them to provide an "opinion as to whether the location of a large scale atomic energy plant in the United Kingdom would be advantageous, disadvantageous, or of limited effect on the security of the United States."[18] Marshall's instructions cut to the heart of the issue. Both British and American officials had already agreed that the one loophole in the McMahon Act was the section that stated that atomic information could be disseminated only if it contributed to the "common defense and security" of the United States.

However, British optimism was short-lived. Eisenhower, who arranged an informal meeting with the Chiefs of Staff to discuss the merit of the British cause, encountered stiff resistance from the military leaders. They saw more security risks than benefits from a British plant. Their March 1947 report to the secretary of state raised three main points that would not be heard for the last time. First, the location of a plant in Britain would place such a plant, and large stocks of usable material, too close to a potential enemy. Canada, it suggested, would be a better location for an atomic pile. Second, construction of a British plant would divert an appreciable portion of raw material from U.S. plants. However, third, the disadvantages of having a plant constructed in the United Kingdom could be minimized if the British plants converted their stockpile of raw material into a form usable for atomic weapons, in which case the fissionable materials would be available to the United States.[19] In other words, the United Kingdom should produce U-235 and plutonium for shipment to the U.S. atomic plants.

To say the least, London was terribly disappointed at this "unfortunate" and "unpalatable" conclusion. The argument was not new. The Chiefs of Staff had raised it in the early months of 1946. At the time, the British had flatly rejected the Canadian location on the grounds that a Canadian plant would not be directly under their physical control. Moreover, neither they nor the Canadians had the financial wherewithal to construct a plant in Canada. With regard to the vulnerability question, the British believed that a nuclear reactor in Britain was in the national and security interest of Britain, since the Americans would be more likely to defend it.[20] Having already conveyed their views, the British were none too happy to hear the issues raised again. With reference to the vulnerability of a British plant, Kennan's Policy Planning Committee (PPC) would take some wind out of that argument late in 1947. An October 1947 PPC report would state that

> the Red Army does not stand on the Channel as did the Wehrmacht. For some time to come the Russians will be in no better position to launch an invasion than the Germans. By the time they do have the ships, the planes and the jump-off points, they will also have the atom plants and the atom bomb. They will not invade Europe and Great Britain merely to learn atomic secrets they already have [the extent of the Soviet spy network was still unknown], nor to seize raw materials which it would take them at least a year to convert into nuclear fuels.[21]

In March 1947, it appeared that the movement toward cooperation would once again be halted, but that was not to be. In October, the United States revived it again. Uranium shortages and other factors forced a reappraisal upon the Americans. Let us trace these new developments.

Beginning in 1947, the State Department and the AEC came together around a common policy on cooperation with the British. The president too became convinced of the merits of collaboration and was forced to walk a tightrope between his policy and that of the congressional guardians of the "secret." The AEC's creation marked Groves's and the military's demise as the controllers of the weapons. Groves had done his job well; it was now time for the emergence of a new group of watchdogs. For the most part this responsibility fell on the JCAE. The AEC had already voiced its concerns that the JCAE had not been informed of the secret agreements, and it was not willing to jeopardize its position by becoming involved in any cover-up. The Senate confirmed Lilienthal as chairman of the AEC on April 9; one week later, he raised the issue at the highest level. At a meeting with Truman, the secretary of the navy, the secretary of war, and Chief of Staff Admiral Leahy, Lilienthal disclosed, to the surprise of the assembled group, that he was aware of the existence of secret "atomic energy agreements during and after the war." He explained that the AEC was concerned that the Senate Foreign Relations Committee had not been informed of the existence of these agreements. The president, supported by Leahy, denied the existence of any agreements. Nonetheless, Lilienthal persisted by adding that it was the understanding of the AEC that a Quebec Agreement existed that "covered not only raw materials and exchange of information but also matters relating to the use of atomic bombs in warfare."[22] The president quickly dropped the matter. He saw no need to include even the chairman of the AEC in that select circle of insiders privy to the wartime agreements.

On a related topic, Lilienthal stressed the importance of maintaining good relations with the British in order to safeguard the supply of uranium. In particular, he mentioned the potential uranium supplies from South Africa, where the United Kingdom exercised some influence—an influence that the United States might be forced to tap in the future in order to gain access to the South African ore. Lilienthal had planted two seeds that would soon take root: one called for good Anglo-American relations and the other for revealing the wartime agreements to Congress. The latter disclosure was already in the pipeline. Acheson had already informed the British that the time was fast approaching when Congress would have to be acquainted with the tripartite arrangements.[23] The American press had started asking questions about secret documents, and a few Republicans were trying to "discover anything discreditable to President Roosevelt or President Truman."[24]

The Republicans did not have to look far or long. Three weeks after his talk with the president, Lilienthal disclosed what he knew to the JCAE. To put it mildly, the members were shocked at the extent of the British and Canadian association with the U.S. bomb project. In his journal Lilienthal reported: "There was some alarm expressed that England is getting half of the Belgian uranium output and some surprise at learning that Great Britain and Canada actually had had men participating with all four feet in the development of the bomb itself, during the war. Senator [Tom] Connally said that then you mean that England knows how to make the bomb. The answer is certainly 'yes.'"[25] This was indeed an announcement of shattering import to men who believed that one of the primary responsibilities of their committee was to prevent the sharing of the "secret" with any other country. Though the British were not fully equipped to build a bomb, on that day the senators may certainly have felt that they had been assigned the task of guarding an empty treasure chest. Having broached the subject, Lilienthal referred the senators to the State Department for more details. This was provided on May 12 when Acheson disclosed the full wartime and postwartime U.K.-U.S.-Canadian agreements to an executive session of the JCAE. They had always believed that the British had made some minor contributions to the development of the bomb. Acheson now told them that the British had contributed much and still had much to contribute and that their scientific and theoretical knowledge of the processes had practically paralleled that of the United States. It is true, they were told, that British industrial "know-how" had lagged far behind that of the United States, but that was due only to the decision of both countries to concentrate development in the United States. Acheson went on to describe the various agreements, including the Quebec Agreement and its articles. One of those articles, the one that stated that the bomb could not be used on third parties without mutual consent, raised the ire of certain members of the committee.[26]

It would be an understatement to describe the senators' reaction to the revelations as one of surprise. Gordon Arneson, special assistant to the secretary of state for atomic energy matters, described their reaction in the following terms: "The hearing room erupted in indignation and anger. Several members walked out at the very thought that we'd have to ask anybody's permission to use these weapons."[27] Three months later, they were still fuming. In a correspondence on the matter to the secretary of state, Senator Bourke Hickenlooper, the Republican chairman of the JCAE from Iowa, stated that he was "shocked and astounded by the information." He believed that the agreement was "ill-advised" and "intolerable" and had to be replaced by a "new and more equitable [to the U.S.] agreement." Hickenlooper was disturbed that the agreement read more like a treaty than a simple memorandum (which it was not). The senator was one of those who frowned upon the United States' getting entangled in foreign commitments. In addition, he considered the 50–50 agreement with the United Kingdom inequitable because the United Kingdom did not need their stockpile and would not be able to produce atomic power for decades. He was "firmly of the opinion" that the British stockpile should be brought to the United States. If the British failed to accede to the U.S. demands, he threatened to "oppose as vigorously as I can, and publicly if necessary, any further aid or assistance to Britain . . . because [the two issues] strike at the heart of our present national security."[28] The State Department's revelation[29] had generated acute ill feeling toward the British and distrust toward the State Department. It would serve to increase the JCAE's vigilance and force it to scrutinize more closely the relationship of both.

Nevertheless, the members of the AEC and State Department "felt much relieved" that the disclosures had been made and that transactions with the United Kingdom and Canada could now be "rather less surreptitious."[30] Carroll Wilson, the AEC's general manager, shared with the British his optimism that a more liberal view of cooperation was imminent. He believed that the disclosures, in light of the failure of international control, now made it easier to persuade the JCAE either to modify the McMahon Act to allow collaboration or to interpret it in favor of collaboration.[31] We have already seen that the State Department too would soon be forced to come to grips with the "grim realities" of the failure of international control and seriously consider rapprochement with the British. The British were pleased to get Wilson's optimistic report so soon after the U.S. Joint Chiefs of Staff had dashed their hopes. They were willing to be patient and not prejudice the chances of a favorable decision by forcing the issue. Besides, the question of cooperation had become less urgent due to a recommendation by the Ministry of Supply that air-cooled piles should be constructed in place of water-cooled piles of the Hanford type. Thus, information on the design and construction of the Hanford type was not critical. Nevertheless, the British still desired specific information on such matters as chemical separation and manipulation of plutonium, which was still of great value and a time-saver.[32]

Three months after Wilson's optimistic report, the British received some more reassuring news, this time from the State Department. A few days after Kennan's policy planning staff had submitted their "grim reality" report, Edmund Gullion, special assistant to Undersecretary of State Lovett for atomic energy matters, informed Donald Maclean, British secretary of the CPC from February 1947 to September 1948, that the State Department was ready to "open discussions regarding co-operation in atomic energy matters."[33] The decision, however, had to be cleared with other government agencies. The British were led to believe that collaboration would be initiated in order to win stronger British support at the UN international control debate in New York. The British embassy summarized this perception as follows: "We are inclined to think that the State Department are in fact cooking plans for closer cooperation with us and the Canadians and have decided to give us a glimpse behind the oven door in order to encourage us to play along with them in New York."[34] This was far from the truth. The Americans had already written off the United Nations and were in the process of formulating a new atomic energy policy that called for a common atomic front against the Soviets. The policy included a tightening of the American monopoly of atomic weapons, an escalated production timetable, and securing of a reliable source of raw material. The British, however, were in a position to adversely affect this policy due to their direct and indirect control of three principal sources of raw material: (1) the stockpile in the United Kingdom, (2) future production in South Africa, and (3) future production from the Belgian Congo. Even a "slowdown" in British cooperation in those areas could have a disastrous effect on the realization of production goals.[35]

There is a tendency to overstate the fact that the Anglophiles—Marshall, Acheson, Lilienthal, Carrol Wilson, and Eisenhower—were well placed in the U.S. atomic energy hierarchy and hence were in a position to address the needs of the British. A note of caution should be injected into this interpretation. All of these men were motivated by the security interest of the United States and not by sentimentality. The Americans' overture in 1947 was influenced by the Americans' growing need for uranium. Moreover, the British did recognize that one of the reasons for the new approach was the Americans' need for raw material. Margaret Gowing stated that "the British did not realize that the main reason for the American anxiety for talks was their determination to get their hands on the uranium which Britain could claim under present arrangements."[36] On the contrary, the British were aware of this and were prepared to resist conceding their uranium to the United States. Eleven days after Gullion had raised the possibility of renewed cooperation, Gordon Munro, who had replaced Makins at the Washington embassy, commented: "It looks as though the Americans are running short of ore, are anxious to keep up production of bombs, and are in consequence developing dishonorable intentions towards our reserves. . . . We expect the question of future shipments, and indeed quite possibly the disposal of UK stocks, to be raised with us."[37] By October 2, Munro had no doubts. Gullion had informed him that unnamed congressmen might raise the question of British uranium reserves in the context of future American financial aid to the United Kingdom. He warned Gullion that "any attempt to put the squeeze on us over our uranium reserves in the context of the Marshall Plan would make the worst possible impression at home and stood no chance at all of success."[38]

What was the state of the U.S. uranium situation at this time? In the fall of 1947, John Gustafson, a mining engineer, became the first director of raw materials for the U.S. program. When he reviewed the raw material records, he was shocked to discover that after a huge outlay of $5 billion in atomic energy expenditure, the entire program "rested on the production of uranium from one mine deep in the Belgian Congo and another small source in the sub-Arctic region of Canada."[39] Congo production for the next four years was projected at 2,000 tons per year, half of which was destined to the United Kingdom; another 150 tons was expected from the Canadian source, near Great Bear Lake.[40] Another Canadian source was the Eldorado company in Canada, which produced uranium as a by-product of its gold-mining venture, but the total amount was limited. Moreover, the Canadian pitchblende was of poor quality, troublesome to mine, and difficult to refine.[41] On the domestic front, an unprofitable and minor source, which was not currently under production, existed in the Colorado Plateau. It was estimated that after extensive renovation the Colorado mines might be able to produce 300 tons per year, but the cost was expected to be prohibitive. For comparison, Congo ore sold for $1.90 a ton for the first 7,500 tons and $3.40 for more inaccessible lower deposits.[42] The Colorado source, with its low uranium content and high development cost, was projected to sell for upwards of $20.00 per ton.[43] The only other known source was in the Union of South Africa, where work was underway to extract uranium as a by-product of the gold-mining operation. It was a great potential source, but the extraction mechanism was not expected to be at full operation until 1950 or so.

Even the Belgian source was beginning to lose its reliability. In 1947 the mine showed signs of exhaustion, raising the likelihood that the mining company would be forced to close it for a year while a new shaft was sunk to exploit new deposits buried deep in the mine.[44] Furthermore, the Belgian cabinet resigned on March 11. The U.S. government feared that the new coalition government, under Paul-Henri Spaak, would nationalize the mines in order to appease the socialist members of the coalition. The State Department speculated that, if the mines were nationalized, the Belgian government would find it difficult to refuse the sale of uranium to the Soviets. Also, they were apprehensive that the communists, who were no longer in the government, would publicly question the special U.K.-U.S.-Belgian agreement, with which they were familiar.[45] In desperation, the U.S. government, in March 1947, had even considered bringing Belgian scientists to the United States for an exchange of information, despite the McMahon Act. It was hoped that this move would lead the Belgian people to assume that Belgium was benefiting from the ore purchase arrangement.[46] The scientists were never brought in, but Prime Minister Spaak was led to believe that they would be allowed to come.[47] In addition, Spaak was continually reassured that the United States would share with Belgium any commercial uses of atomic energy that might be discovered in the future. It was this reassurance that reconciled him to the cheap price the United States paid for the Congo ores. Spaak was never informed that the McMahon Act had invalidated any such prior agreement,[48] and if he knew he did not say so. This illustrates how far the United States was willing to go to procure raw materials.

The British found out about the American ore shortage problem in September. The JCAE already knew about the shortages, having been informed by Lilienthal in May. Lilienthal reported that some members of the JCAE had actually believed that U.S. domestic sources would be sufficient for U.S. needs[49] and stated that it took some doing to convince them that the United States faced a shortage crisis.[50] As the year progressed, the uranium deficiency problem became a question of growing concern for the AEC, the Joint Chiefs of Staff, and administration officials. They all knew that the solution to the problem was the British stockpile and that steps had to be devised to gain possession of it. The AEC prodded the departments into action on October 1, 1947. In a letter to Kennan's policy planning staff, it discussed officially the shortage problem. The AEC pointed out that the joint chiefs believed that if the atomic bomb program were reduced, the impact on national security would be adverse. Thus, national security required the expansion and preemption of the world production of uranium. The quantity of raw materials needed to see the United States through the critical period of shortage could be obtained only from the Belgian Congo, the U.K. stockpile, and, beginning in 1950, South Africa.[51]

Kennan and Gullion, in turn, submitted a memorandum to the new undersecretary of state, Robert Lovett—Acheson had stepped down in July. (Lovett was a former assistant secretary of war; he too would support the renewal of atomic relations with the British.) They argued that in view of the shortages of raw materials it was in the interest of the United States to improve atomic energy relations with the British and Canadians. In addition, the JCAE and the two congressional foreign relations committees should be apprised of the necessity of sharing atomic information with the United Kingdom and Canada. Once the congressional leaders were won over, the two countries should be invited to informal secret talks to discuss, ostensibly, the failure of the UN Atomic Energy Commission and "our future dealings with one another in the procurement of materials and in the exchange of information."[52] Thus was born the new U.S. initiative to improve relations with the British. This initiative was taken not because of any newfound desire to assist the British in the development of their atomic program but because the national interest and security of the United States was in dire need of British help.

As a follow-up to Kennan's memorandum, several high-level meetings were held over the following weeks, attended by, among others, Secretary of State Marshall, Secretary of Defense Forrestal, Secretary of the Army Kenneth Royall, Lilienthal, Kennan, and Dr. Vannevar Bush.[53] The consensus was that the United States could benefit from an exchange of information with the British. Moreover, the United States could get around the McMahon Act by convincing the JCAE that the interchange was in the interest of national defense as provided for in the act. Lilienthal advised that when the JCAE was approached emphasis should be placed upon the advantages to the United States of cooperation in order to win over its members.[54] Steps were quickly taken to win congressional approval. On November 26, a meeting was held at Blair House, to which Senators Hickenlooper and Vandenberg were invited. (In addition to his duties as a member of the JCAE, Vandenberg also served as chairman of the Senate Foreign Relations Committee.) Lovett summarized the U.S. goals to the senators as (1) "tidying up" the wartime agreements; (2) dispersing the stockpile of raw materials in Britain; (3) getting a satisfactory share of Belgian production; (4) restricting raw material storage in Britain to the amount which could be used in their current projects; and (5) obtaining British and Canadian support in negotiations with South Africa for their uranium.[55] The senators raised no objections to the administration's goals. The proposal also included the exchange of atomic technical information with the British. Lilienthal and Lovett went to the meeting anticipating stiff opposition from the senators for the latter proposal; however, to Lilienthal's surprise, "it was received almost without question, in every essential, the thing we thought would cause the most trouble didn't raise a ripple hardly."[56]

Nevertheless, Senator Vandenberg repeated something he had said weeks before. He stated that no arrangement would be acceptable that obligated the United States to seek the consent of any country before using the bomb. "The provision in the wartime agreements to this effect absolutely had to go."[57] He repeated the earlier concerns of Hickenlooper that it was inadmissible for the United Kingdom to stockpile uranium that the United States needed. He threatened that if the "hoarding problem" were not settled by the time the Economic Recovery Program came before Congress in the next three weeks, then he "would certainly see to it that any future legislation or further loan assistance to the British took account of their failure to meet us at least halfway."[58] Lovett, Kennan, and Royall were vehemently opposed to using any such threat. As a matter of fact, an internal State Department report had already concluded that the Marshall Plan aid should not be tied to the acquisition of uranium. It read:

> If our aid to Britain . . . is a good thing, it is a good thing regardless of whether we gain uranium thereby. The amount we spend on European reconstruction is really an insurance premium on the continued existence of the kind of world economy in which this country is the principal shareholder. We have more to gain by its survival and more to lose by its death than any other country. . . . Every Communist party in Europe would gain ground if it could say that we were bargaining relief of human suffering against a perpetual US monopoly on atomic bombs.[59]

At the Blair House session, Lovett calmly explained, for the benefit of the senators, that he did not believe that it would be necessary to use threats at the outset of negotiations. He believed that the United States would gain what they wanted through normal diplomatic channels. "There was no harm, however, in keeping the 'big stick' in plain sight in the corner, even if we gave no indication of any immediate disposition to use it."[60] The senators and the administration representatives left the meeting fully satisfied. The next step called for getting the Canadians and British to Washington.

How well did Lovett use the "big stick"? He used it quite effectively. Gullion had already forewarned Munro that certain senators were considering tying Marshall aid to uranium supplies. On November 13 the warning came to pass. Senator Vandenberg raised the issue in the course of hearings by the Senate Foreign Relations Committee.[61] Lovett, although he disagreed with the senator's approach, could not have asked for a more opportune development. He seized upon the publicity it generated to impress upon the British that he was about to be called before the Senate to explain why the two issues should not be connected. If he were called, he explained, the whole story of the wartime relationship would become public knowledge, which would in turn lead to an endless congressional wrangle. One way around the problem was to work out immediately a new basis of cooperation that would support his contention before the Senate that the matter was well in hand and should, on national security grounds, not be dragged into the Marshall debate. Lovett then told Munro that he was taking the risk of saying as much as he had "in order to give you some advance notice of what is in the wind."[62] The following day Lovett told the British ambassador that any new atomic energy discussions would have to be completed by December 17, which was the very latest date to which he could postpone his interrogation.[63]

It is noteworthy that Lovett, at these exploratory meetings, did not dwell upon the U.K. stockpile as it related to U.S. shortages. Rather, in response to a British "feeler about raw materials," he hinted, "rather guardedly, at the military temptation offered to the Soviet Union by a stockpile in the UK."[64] Over the next few days Lovett continued to keep the "big stick in plain sight." On December 5, the full JCAE consented to the holding of Anglo-American-Canadian discussions. The following day, Lovett cabled Marshall, who was attending a meeting in London, asking him to discuss the new U.S. initiative with Bevin, the British foreign secretary. He asked Marshall to emphasize the importance of concluding the negotiations quickly before Congress tied them to the European Recovery Program.[65] A week later, the U.S. ambassador to the United Kingdom informed Lovett that he had conveyed the same message to Makins before Makins's departure to attend the Washington discussions.[66] In the same telegram he added that he had also spoken to Bevin "and briefly discussed the matter explaining our concern that the subject be not involved in ERP debates."[67]

The atmosphere surrounding the preparation for and the actual negotiations was charged with an acute sense of urgency. The JCAE approval came on Friday, December 8, and the first preliminary meeting of the CPC was summoned for that Monday. A frantic scramble ensued in London as the British called an impromptu cabinet meeting to consider the brief for the discussions. They quickly put together a negotiating team and hastily booked air passage and hotel accommodations. Lord Inverchapel commented: "We recognize that this sort of rush after two years delay seems rather absurd."[68] Two days later, with the British team not due to arrive for two more days, Makins too commented: "I am sure Americans will understand that we are not REPEAT not in any way hanging back but that they must give us at least two days to consider issues which they have been mulling over for two years."[69] One may assume that the British team was not fully rehearsed for these discussions, considering the atmosphere of urgency the Americans had created and the haste to get a British team to Washington. Fortunately, however, the two senior members of the assembled team, Cockcroft and Makins, brought with them several years of intimate involvement with atomic energy problems. As head of the U.K. program, Cockcroft was familiar with the scientific and technical needs of the atomic energy establishment. Makins, on the other hand, was fully versed in all the diplomatic setbacks.

Moreover, despite Lovett's attempt to downplay the allocation issue, the British were not fooled. They knew that the Americans were experiencing shortages and would seek all of the Congo shipments and, quite likely, the stockpile lying idle in Britain. Hence, they drew up their own negotiating stance with that recognition in mind. Makins described the cabinet's negotiating position in the following terms: "We regard it essential that exchange of information should have at least as firm a place on the agenda as allocation of raw materials. Our objective is a comprehensive agreement on the future basis of collaboration, and our agreement on allocation would be dependent on a satisfactory agreement on exchange of information."[70] Further, the cabinet was determined not to yield an ounce of its stockpile. Its instructions to the negotiators included the emphatic statement: "Above all it is essential not to yield to any American demand to transfer some of our existing substantial stocks of uranium ore out of this country."[71] Hence, both teams met in Washington with conflicting priorities. The Americans wanted to gain possession of a stockpile that the British were unwilling to yield. On their side, the British sought full atomic collaboration, which the Americans could not and would not concede.

The United States rolled out the red carpet for the British delegation. Makins's initial report to London was optimistic: "The atmosphere in Washington was friendly. . . . There was every sign that the Americans wanted to reach a fresh understanding so that in the future all three governments would play from the same score."[72] Cockcroft was equally impressed with the willingness of members of the AEC to help. Guy Hartcup and T. E. Allibone, Cockcroft's biographers, explain that "at a supper party . . . good relations were strengthened as they sat in front of a log fire and 'ate cold chicken and rice.'"[73] At the first meeting of the CPC, Lovett informed the British that he hoped that their conversations would mark the resumption of regular, friendly, and informal contact among the three governments and that the United States "aimed at a resumption and reaffirmation of the close association we had in the past in matters of concern to the Committee."[74] Obviously, the British were impressed by these friendly overtures and were not willing to do anything to disturb the friendly atmosphere.

The CPC appointed the Subgroup on Technical Cooperation, which included one representative each from the United Kingdom and Canada and four from the United States. This subgroup submitted nine areas in which cooperation would be "mutually advantageous." The list disappointed the British, primarily because it excluded the technological and military questions that interested them; yet they protested weakly. Included on the list were three areas that were already declassified and open to all: (a) subjects covered in sections I and II of the Proposed Declassification Guide of 1947; (b) the entire field of health and safety; and (c) general research experience with the low power reactors named in (a). Also included were two areas of technical importance where British research was just as advanced as, and in some cases even more advanced than, the United States.[75] The two areas were extraction of low-grade ores (g) and the design of natural uranium reactors (h). Additional areas included research uses of radioactive isotopes and stable isotopes (e); detection of a distant nuclear explosion (d); survey methods for source materials (e); and beneficiation of ores (f)—cooperation with South Africa and other dominions. The British accepted the list with a vague American assurance that more topics could be added to the list in the future.

Getting the British to make concessions was not too difficult. The British delegation seemed determined not to muddy the waters of cooperation by being too inflexible. Only when it was asked to surrender the stockpile did the delegation harden its resistance. It was willing to surrender the entire Congo production for the next year or two but not the stockpile. By December 18, agreement in principle had been reached on all points with the exception of allocation. Both teams agreed that Makins, Cockcroft, and David Peirson should return to London for further instructions.[76] Kennan was convinced, and rightly so, that the stalemate stemmed not from the delegation itself but from the cabinet in London, where the decision had already been made, largely on emotional grounds, as he put it, to stand fast.[77] He solicited the help of Marshall, who was still in London, to persuade Bevin to use his influence on the cabinet. Meanwhile, an effective argument was made to the British delegates. Despite its firm instructions from London, the British team accepted the persuasive American argument that the stockpile should be released to feed the U.S. atomic plants, since, in Lovett's words, "no part of the Western world could feel secure if the United States bomb production program was to be handicapped."[78] Makins took the argument back with him to London to get cabinet authorization. No doubt, he also added that "Lilienthal and Bush [had] presented the nine areas of agreement as but a beginning, [that] Lovett had spoken of cooperation as a continuing effort, and [that even] Forrestal had described the three nations as partners."[79]

The inner circle of the cabinet was persuaded, and the needed concession was given its grudging approval. A new agreement was signed on January 7, 1948. The agreement provided that

1. All wartime agreements were null and void, except those articles that related to raw materials. There was no further need to get the "consent" of the United Kingdom before using the bomb. (However, from 1948 to 1953 the British would attempt to gain the right of "consultation" before the use of U.S. bombers from U.K. airfields.)
2. All supplies of uranium produced in the Congo were to be allocated to the United States in 1948 and 1949.

3. If the United States needed additional raw materials to maintain its minimum program, it would be provided from the British stockpile.
4. The exchange of atomic information would be shared in the nine aforementioned mutually agreed areas.[80]

Gullion suggested that the agreement be called Modus Vivendi, a term most often used to describe the relations between adversaries driven by circumstances to get along together. He thought that the term was most accurate.[81]

The British were not exactly ecstatic about the agreement; it fell far short of their expectations. Why, then, did they accept it? Cockcroft would later rationalize that it was "not exactly an exciting result, but at least it was a beginning."[82] He was referring to the beginning of postwar atomic collaboration. He anticipated that the modest concessions made by the United States would improve generously with time. The U.S. officials had encouraged that line of thought. In the mind of Bevin, the agreement also marked a beginning, the inception of a broad range of Anglo-American international cooperation. Bevin was not willing to trifle with the Marshall Plan or risk losing it. He saw it as the salvation for the economic woes of Britain and Europe. From the delivery of Marshall's speech at Harvard on June 5, 1947, Bevin embraced its reconstructive potential, coming as it did at a time of convertibility of sterling, unemployment, and the recent fuel shortage experience. Bevin, therefore, could not risk inciting the Americans to bring the "big stick" into the negotiating arena. Neither, for that matter, could Attlee and the cabinet gamble with the economic aid if they expected to realize the full implementation of Labour's socialist programs. Capitalist America could once again, as it had with the 1946 loan, assist Labour in assembling the framework of socialism. Making atomic energy concessions was a small price to pay, especially since there were no immediate demands for uranium in Britain.

Furthermore, one cannot lose sight of the Soviet threat. Before the Berlin crisis of 1948, Bevin did not view the Soviets as a threat to the security of Britain, but he did perceive them as a serious threat to the peace and security of Europe. In June 1947, both he and Attlee had frowned upon a Chiefs of Staff report that based British defense policy on the possibility of war with Russia.[83] They were not alone. Sir Henry Tizard agreed that a British war with Russia was remote. He argued that "in spite of aggressive words in public, there is no indication the Russian rulers have any intention of risking a major war in the near future."[84] Nevertheless, Bevin did view the Soviets with grave suspicion. He admitted that they planned to expand in a piecemeal fashion. He also believed that they were intransigent and that it was best to force public confrontations against them in the Security Council so that the world could see them for what they were.[85] By 1947 no love was lost between Bevin and the Soviet leadership. As early as the first Council of Foreign Ministers meeting in London, relations between him and Molotov got off on the wrong foot as the Russian questioned the British brand of socialism. The tension continued to build as Soviet intentions in Europe became clear. At one conference, at the end of 1946, Bevin, tired of hearing Molotov defend Soviet policies and attack British proposals, charged at Molotov with clenched fists, only to be restrained by security guards.[86] The last straw was the failure of the last Council of Foreign Ministers Conference that was held in New York in December 1947. This meeting crystallized the irreconcilable differences between the Soviet Union and the Western powers in their perception of Europe. It seems that the meeting convinced Bevin that the security of Europe could be safeguarded against Soviet expansionism only thorough the creation of some sort of European security zone that included the United States. He had already convinced himself that European economic viability was not possible without the injection of American dollars.[87] On December 17 and 18, dates that coincided with the Washington stalemate over allocation, Bevin outlined to French Foreign Minister Georges Bidault and Marshall his vision of a "Western Union." With Marshall's enthusiastic blessing, he and the leaders of France, Belgium, the Netherlands, and Luxembourg set about organizing the union. The Commons and the public were officially informed of the idea on January 22, 1948, in a speech in which Bevin summarized the expansionism and inflexibility of the Soviet Union and laid stress on the political, economic, and spiritual unity of Western Europe.[88] This union would take shape with the signing of the Treaty of Brussels on March 17, a mutual collective self-defense treaty that, though not naming a potential aggressor, was widely interpreted as being directed at the Soviet Union.

This treaty, then, was under quiet discussion at the time of the atomic energy negotiations. It is quite possible that as much as Bevin was determined not to jeopardize the Marshall Plan aid, he was equally prepared not to abort his embryonic union by being stingy with the atomic stockpile. He could not risk alienating the United States, for without its support no anti-Soviet alliance was worth even an ounce of uranium ore. Bevin left nothing to chance in his efforts to get the United States to commit itself to the security of Europe. From the outset he sought Marshall's blessing and kept him fully informed of developments. Moreover, on the diplomatic front, Sir Oliver Franks, the new ambassador to the United States, was used to persuade the United States to abandon its reluctance to commit troops permanently to Europe. Furthermore, five days after the signing of the Brussels Treaty, a powerful British delegation, under Gladwyn Jebb, was sent to Washington. This delegation contributed to the evolution of the idea of the North Atlantic Treaty Organization.[89] Any attempt to understand the rationale behind the United Kingdom's decision to concede as much as it did for as little as it received must take into consideration the larger context of British foreign and domestic policies during this period. One cannot look at these negotiations in isolation from other issues. For the most part, during the period under study, both governments did keep the issue of atomic energy separate from other international matters. On this occasion, however, the American negotiators interjected the Marshall Plan, and Bevin allowed himself to be influenced by factors other than atomic energy.

The U.S. negotiators were completely satisfied with the Modus Vivendi. In an address to the JCAE, Lovett, in glowing terms, expressed his satisfaction: "In fact, I think we have achieved more than we might have expected before the talks were begun."[90] Senators Vandenberg and Hickenlooper, at an earlier briefing, had already expressed similar sentiments. Vandenberg confessed that more had been achieved than he thought possible; he added that "the State Department negotiations represented a considerable accomplishment."[91] He also expressed relief that the wartime agreements had been rescinded. In reference to the interchange of information, Lovett had allayed any misgivings the JCAE might have had by explaining that the proposed exchange was by no means disadvantageous to the United States: "In fact it seems to be heavily loaded in our favor. We stand to learn more than we give. The criterion for exchange of information will be the degree to which such an interchange would promote the national security of this country in the terms of the Atomic Energy Act of 1946."[92] Thus, long before the ink on the Modus Vivendi had dried, the United States was talking in terms of less than full collaboration—at least to the JCAE.

In the course of discussions with the British, the United States still led them to expect more than the Modus promised. One week after the agreement was signed, Kennan made it clear to Makins that the State Department "intended to stand no nonsense" from anyone who stood in the way of its implementation. If need be, he was prepared to recommend amending the McMahon Act or recommend replacing one or two commissioners if they gave trouble.[93] The British were confident that the State Department would stand behind the agreement. They were also aware that the agreement provided for exchanges only in "mutually agreed areas." Dr. F. Woodward, the scientific attaché at the British embassy, admitted several months later that it was generally understood at the time that it was not the purpose of the Modus "to exchange technical information which would convey information to the UK on the manufacture of weapons."[94] In London the Labour government did not attempt to use the agreement to win badly needed public support; it could foresee only public and parliamentary criticism. The cabinet strongly resisted any attempt to publish the results of the talks. It decided that the Modus should be classified as "Top Secret."[95]

The Modus Vivendi turned out to be, not the new beginning that the British envisioned, but the continuation of the steady decline of Anglo-American postwar collaboration. The execution of its terms soon led to more misunderstandings and misinterpretations, to be described in the next chapter. However, despite the problems that emerged during the execution stage of the agreement, it does seem that when the State Department set about to negotiate the Modus it was sincere in its effort to improve atomic relations with the British. It is also true that it was fully prepared to make concessions because it recognized that cooperation served the U.S. security interest. Independently, the AEC also came to the same conclusion, but the commission exercised a little more restraint in terms of the lengths to which it was willing to go to win over the British. Although the United States was running short of uranium, Lilienthal and Lewis Strauss, another commissioner, were not willing to go as far as seeking an amendment to the McMahon Act, as the State Department had suggested, to facilitate collaboration. Lilienthal thought that they could work within or around the constraints of the existing law. To both the State Department's and the AEC's credit, they won over the JCAE. However, that support came with the distasteful rider that tied the negotiations to the Marshall Plan. Senators Hickenlooper and Vandenberg thrust upon the State Department the idea of using economic aid to wring concessions from the British. As unpleasant as the plan was perceived, the State Department astutely executed it. In the end, the British accepted the terms of the Modus due to a combination of American pressure, Soviet intransigence, and British domestic and foreign considerations. If the Americans were soon talking in terms of less than full collaboration, it was because the Modus Vivendi provided for minimal cooperation. The British knew this from the outset. Yet, having allowed themselves to be swayed by the vague promises and honeyed words of the U.S. negotiators, the British, in the immediate future, came to read more than was actually stated. The years 1948 and 1949 would be filled with more of the same frustrations experienced since the end of the war.

SEVEN

Congress Derails the
Modus Vivendi

The Modus Vivendi worked relatively well for six months. During that period its course was fraught with some disappointments and delays, but for the most part the British were satisfied.[1] The delays were caused not by a reluctance on the part of the Atomic Energy Commission (AEC) to exchange information under the nine agreed-on areas but rather by the cumbersome bureaucratic procedures set up to clear each topic proposed for interchange. As a result, the British did lodge mild protests from time to time. Examples of delays were well documented. For instance, on several occasions British technical teams had to postpone their departure from London or, after arriving in the United States, change their itinerary at short notice while waiting for AEC approval to visit installations or meet with American scientists. Despite their protests, the British understood that the difficulties were due mainly to the extreme caution the commission felt was needed in the early days of collaboration. By June 1948, Cockcroft was satisfied that the Americans were holding little back.[2] They were cooperating fully in the nine agreed-on areas. In fact, he thought that the time was opportune to ask U.S. officials to consider, as they had promised, the inclusion of additional areas.[3] Good relations, however, were not destined to last. A series of developments would once again restrict the scope of collaboration.

The first centered on the British decision to reveal the existence of its bomb project. Secrecy had created its own problems. The program had expanded so rapidly that the government found itself faced with an ever-growing risk of leakage to the press. Moreover, both senior and junior officials who were not in the know interpreted their exclusion as an apparent lack of trust. A further reason was that the research establishment experienced serious problems in recruiting scientists whom it could not tell what they were wanted for.[4] Indeed, one could see why that would be a problem. The Chiefs of Staff and the prime minister endorsed Lord Portal's recommendation that the project be made public. On May 14, the minister of defense, in response to a planted question, announced to the House of Commons that the government was developing atomic weapons.[5] Finally, after eighteen months of evasion and denials, the existence of the project was out in the open. Despite this admission, secrecy was still maintained. As a precaution against undue publicity and unwanted revelations, the government gave the project a "D" notice, which, in effect, prevented the press from publishing "any information or speculation as to technical details, the location of research, the persons engaged in it or the time factor."[6]

Before its public announcement the government also considered the problem of American reaction. To date, the British had not notified the Americans officially that they were engaged in a weapons project, only in atomic research. Steps had to be taken to address that issue before going public. On March 19, Donald Maclean, acting on instructions from London, informed Edmund Gullion, special assistant to Undersecretary of State Lovett, that the United Kingdom "had been engaged in research and development work in atomic weapons."[7] Two weeks later, Secretary of Defense Forrestal received similar notification from Admiral Sir Henry Moore. Neither the State Department nor the secretary of defense expressed any concerns. The British had assumed, rightly, that the Americans already knew about their bomb project and that the Americans had chosen not to ask sensitive questions about it. It was one of those subjects that was best left unbroached. Indeed, Forrestal was surprised "that the existence of such work was not already publicly known in the United Kingdom as it was in the United States."[8]

But were the Joint Congressional Committee for Atomic Energy (JCAE) and the AEC aware of the existence of the British project? The British announcement did not seem to strike a chord among members of the AEC and the JCAE. That is, they did not react, one way or the other, to the British disclosure. Subsequent events would indicate, however, that the British bomb program was not as widely known among those who should have been kept informed as the general silence would lead us to believe. It would also become clear that when the State Department and Forrestal received their "official" notification, this information was not passed on. The Modus would begin unraveling within weeks of the public disclosure.

The origin of the first controversy was not directly related to the parliamentary disclosure. In May and June, the British hosted an AEC scientific team of Walter Zinn of Argonne Laboratories, George Weil of the reactor branch of the AEC, and C. W. J. Wende of Hanford. By all reports the British rolled out their red carpet and gave the Americans access to their most secret operations, including their reactor plant for the production of plutonium at Risley. It was the first time any Americans had seen the British plutonium plant, and the confirmation of its existence caused quite a stir among the AEC scientists. Since they were under the impression that collaboration was restricted to non-weapon areas, they refused to disclose any bomb-producing information requested by the British.

Upon their return to the United States, the team reported to Lilienthal that its members were "surprised" and "disturbed" to learn that the British were engaged in more than basic research.[9] Lilienthal quickly checked his minutes and discussions leading up to the Modus and concluded that the agreement did not forbid the construction of British bomb production plants. However, he was "sure, reasonably sure, no one thought that this would be in the British plans."[10] The following day, July 6, he informed the American members of the Combined Policy Committee (CPC) of the team's report. He found little sympathy for his concerns. Lovett explained that although the question of plutonium production had not been raised in the December–January negotiations, it had not been assumed that the British would not produce weapons. Furthermore, he reminded the members, the United States was benefiting "handsomely" in the field of raw material, and the AEC's finding "should not in any way affect existing arrangements."[11] Donald Carpenter, chairman of the Military Liaison Committee, expressed similar sentiments. The consensus of the committee was that cooperation should continue in the agreed-on nine areas but that the United States should not initiate steps to add new areas. Revealingly, the CPC also agreed that if the British made a formal attempt to add new areas, the committee should seriously consider doing so.[12] If nothing else, the meeting seemed to crystallize the thinking of the State Department, the Military Liaison Committee, and Lilienthal on the issue of broader collaboration. They were not willing to risk offending the British, thereby jeopardizing the raw material situation.

Not every member of the AEC took the news of the U.K. production plants with a shrug. Commissioner Lewis Strauss took it very emotionally and suggested to Lilienthal that the president be informed. Lilienthal wrote in his journal that during their conversation Strauss's hand trembled most of the time. Lilienthal surmised that this was so because Strauss felt that "Britain is 'far to our left' and therefore may give 'the secret' away to the Communists, 'some of whom actually sit in Parliament.'" Later that day Lovett informed Lilienthal that Strauss was "very active all over the place," criticizing the British action. Lovett was afraid that the British would find out and conclude that the United States once more was "planning to let them down."[13] He asked Lilienthal to get Forrestal and Bush to "straighten Lewis out." Forrestal in turn sought the advice of Bush, a member of the U.S. negotiating team, who assured him that the British had not failed to abide by the terms of the Modus. The American members of the CPC, he explained, had not expected the British to "refrain from either production of plutonium or the development of atomic weapons." Bush was "somewhat at a loss to understand how on this record any question could now be raised as to the intentions of the British Government."[14] Strauss, like Lilienthal before him, found no sympathy for his concerns. Unlike his chairman, however, he broke ranks. This lack of support led Strauss to seek out other avenues to voice his disapproval. He allied himself with Senators Hickenlooper and Vandenberg, and together these men would oppose any move to expand the range of topics to be exchanged. It was not long before this alliance was given an opportunity to show its solidarity.

The storm broke in July–August 1948 in the so-called "Cyril Smith Affair." The effect of this affair muddied the waters of collaboration and ended any effort to expand upon the nine areas. At the end of July, Dr. J. B. Fisk, the AEC director of research, sent Dr. Cyril Smith, a metallurgist and member of the AEC's General Advisory Committee, to England to discuss a list of subjects. Two weeks after he left, Rear Admiral John Gingrich, the AEC chief of security, got hold of the list, where he noticed that one of the items of discussion was "the basic metallurgy of plutonium." He apprised Strauss of his observation; Strauss in turn informed Sumner Pike, acting chairman of the AEC—Lilienthal was out of town. According to Forrestal, Strauss thought that Pike was too slow to act, so he took the matter to Senators Hickenlooper and Vandenberg.[15] They agreed that the Modus Vivendi did not call for the exchange of bomb information. They went to Forrestal, who called in Bush and Carpenter. Hickenlooper argued that the exchange of information on the "basic metallurgy of plutonium" was definitely weapons information and that there was no possible justification for its exchange under the Modus.[16] Bush and Carpenter agreed. Vandenberg urged Carpenter, as chairman of the Military Liaison Committee, to exercise his authority and monitor the exchange of information more closely.[17]

Carpenter relayed the objections to Pike, who informed him that he had already sent two cables to Cyril Smith instructing him to drop the offending item from the list. Smith was on a tour of Scotland, but he did receive the cables before any exchanges had taken place. As a result of the "Cyril Smith Affair," the Military Liaison Committee, in late August, included in all its instructions to its representatives the following statement: "While recognizing that a distinction between atomic energy matters of military significance and non-military significance cannot be clearly made, all exchanges shall be further governed by the general criterion that information specifically relating to weapons or to the design or operation of present plants for production of weapons materials or weapons parts is not subject for discussion."[18] This episode ended any movement toward including bomb information under the umbrella of the Modus. It would take the British several more months to realize that any such expectation was futile.

The Modus was an intentionally limited document, which the State Department and the AEC had hoped to work around without offending the JCAE. It appears that the State Department restricted the scope of the exchanges, on paper, to deceive the JCAE and not to shortchange the British. The members of the JCAE were not present at the negotiating sessions; hence, they were not privy to what was discussed or what was assumed. Their information came from the overly optimistic and even gloating report of the administration in January. As a matter of fact, it is quite clear that Hickenlooper had not read the agreement; it is uncertain whether it was made available to him and other members of the JCAE. At the meeting with Forrestal, Carpenter, and Bush on August 12, Hickenlooper stated that at the time the cooperation plan was discussed with him and Vandenberg, the previous December, "it was understood that information would be exchanged in 3 areas," not the nine areas now being exchanged, and "that England's primary activity was to be along the lines of power production, and there was no indication of their entering into weapon production."[19]

The report concentrated on what was gained and not on what the United States had tacitly agreed to concede: acquiescence to the existence of a bomb program in England. A revealing memo by Makins to Cockcroft throws some light on the issue of deception. He wrote: "You will remember that during the negotiations in January we were given the tip not to press for an exchange on this topic [weapons information] as the U.S. side were not ready to agree. Ministers therefore decided to leave the question on one side but to raise it later on when collaboration had made some progress."[20] That is, they would keep weapons information out of the agreement in order to gain the support of the JCAE, but once collaboration took off and Congress was lulled into a sense of complacency, they would agree to expand the scope of the agreement. Is it possible that this anonymous "tipster" was a State Department official and that the British had aired their future plans to the Americans around that time? If so, that would partly explain why the State Department was not taken aback by subsequent revelations surrounding the British bomb program.

Did the State Department keep the AEC informed about the British plans? When one considers Strauss's agitation, the surprise of the technical team that "discovered" the plutonium plant, and Lilienthal's initial reaction to the plant, the answer is no. Nonetheless, after Lilienthal learned about the British weapons program in early July, he too seemed to acquiesce to avoid adversely affecting his raw material pipeline. The "Cyril Smith Affair" is an example of this acquiescence. Smith was sent to England after the plutonium plant was discovered. Yet Pike, probably with Lilienthal's blessing, approved the exchange of information that went beyond the scope of the agreement. The consensus was that the Modus did not provide for the exchange of metallurgical information. Was its inclusion an oversight or a loose interpretation of the Modus? Carpenter too noticed that the extent of the contacts had "started out rather gradually and has just recently become so extensive as to require definite control procedures."[21]

The contacts had started to expand because Lilienthal was willing to make concessions to keep the British happy. He was quite displeased with Carpenter's new restrictive guidelines. He was afraid that exchanges might become "so niggardly and reluctant" as to subvert the spirit and intent of the agreement and in the process affect the raw materials situation. He had reasons to be concerned; the British might have to be contacted again concerning another allocation arrangement for 1950 and 1951. Their response, he warned, might not then be favorable.[22] Lilienthal was not alone in his apprehension. Gordon Arneson also reminded Lovett that negotiations with the South Africans were soon to begin and that British support would be needed. In view of the high raw materials stake involved—and the question of honoring the spirit and letter of the agreement—Arneson suggested that the State Department support a more relaxed method of cooperation.[23] Despite the efforts of the AEC and the State Department, the "Cyril Smith Affair" and the revelation of the British plutonium reactor had left its mark on present and future collaboration efforts.

The British, not sensing what was happening in the United States, took steps that did not help matters any. It will be recalled that in June Cockcroft had stated that he thought the time was opportune to seek an extension of the nine areas. Makins agreed. He saw some advantage in making the request as soon as Congress adjourned at the end of June, since the members of the JCAE would be dispersed and, it was hoped, distracted. At the latest, it should be made before the presidential and congressional elections of November. He was afraid that those officials responsible for drawing up the agreement might not return to Washington.[24] This was a real fear, since the polls indicated that Truman faced an uphill fight to retain office. By August 25, the prime minister, foreign secretary, minister of defense, and Chiefs of Staff all agreed to make a formal request for increased collaboration.[25]

The decision was not timely. On August 16, four days after the Military Liaison Committee had drawn up its guidelines, Carpenter informed Dr.

Woodward, the scientific attaché to the British embassy, that the JCAE was considerably concerned about the British plutonium program. As a result, he pointed out, exchanges within the nine areas would have to be carefully "policed," and expansion of exchanges was quite improbable.[26] Dr. Woodward, strongly supported by the British ambassador, conveyed the information and their support for it to London. The prime minister overrode them, citing the previously mentioned factors that Makins had raised. Moreover, Attlee accepted the additional recommendation that the request be submitted, not through the AEC or State Department, but directly to the military establishment.[27] No doubt, London hoped that the military men would better appreciate the security arguments raised to justify the exchange of bomb information. This turned out to be a mistaken assumption. Making the request at that time also turned out to be ill advised.

On September 2, Admiral Sir Henry Moore, head of the British Naval Mission in the United States, presented Forrestal with a formal request from the British minister of defense, A. V. Alexander. The memo argued that the development of the British atomic program would contribute to the security of both nations, since the British would be better able to help the United States maintain world peace.[28] The timing for such a security argument seemed appropriate, since the British and the Americans were working together to resolve the Berlin crisis. Nevertheless, Forrestal advised Admiral Moore that the timing "of the request" was inappropriate and that it would be unwise to press for a prompt answer. All the same, he sent the request to the Chiefs of Staff for their opinion. Carpenter was livid that the British had not heeded his advice. On September 16, he indelicately voiced his displeasure to Woodward. He raised the issue of British security and American concerns that British sources might leak atomic information to Moscow. Woodward did not take too kindly to that insinuation, even countering that U.K. security was better than that of the United States. Carpenter then suggested that the United Kingdom consider discontinuing the manufacture of weapons altogether in exchange for a few atomic bombs from the United States for use by the United Kingdom in an emergency.[29] (Woodward did not endorse this idea, but it would resurface again and be explored in the future.) Predictably, on September 29, the Joint Chiefs of Staff went formally on record as strongly opposed to any extension of the areas of exchange.[30]

Attlee and his people in London had erred in their judgment on several counts. First, they should have heeded the advice of Carpenter, transmitted through Woodward, to refrain from making demands for an extension of the nine areas. The British ambassador, who was in close proximity to the pulse of the U.S. government, had also suggested restraint. All things considered, the climate was not right. It is true that London's attention was focused on the forthcoming elections and the possible changing of the guard, but the British should have been equally cognizant that the "watchdogs" were ever more vigilant in their efforts to protect the "secret." Not until September 20, after Woodward had had his confrontation with Carpenter, did it dawn on him that the "tightening up" on the part of the United States might have been due to the visit of Zinn and Weil in June. He felt that London had erred in showing them the plutonium plant.[31] Second, by sending the request through military channels and not through the State Department or AEC, the British had selected the worst avenue possible. Outside of the JCAE, no group supported closer collaboration less. The defense establishment tended toward a very restrictive interpretation of the nine areas so as to deprive the British of bomb-producing information. Hence, they were not likely to embrace the British proposal for an extension of the areas of exchange.

Channeling the request through the AEC might have been a more worthwhile gamble. In general, the AEC tended to be more liberal in its interpretation of the nine areas. We have already discussed Lilienthal's concerns of November 2 that the exchange of information was becoming too "niggardly." The bottom line for the AEC was always raw material. One of the most important functions of the AEC was to produce atomic weapons, and that could not be done without raw materials. Hence, the AEC, with the exception of Strauss, was more willing to be flexible by sacrificing some atomic information in order to safeguard its source of raw material. This is not to say that the British would have won an automatic extension if they had appealed to the AEC. The AEC had to weigh the views of the Military Liaison Committee and the JCAE, the two "watchdogs" created by the Atomic Energy Act. But the AEC might have more readily accepted and projected the mutual security argument.

One day after the Joint Chiefs of Staff decision, Sir Oliver Franks, the new U.K. ambassador, lodged a protest with Lovett from the Official Committee on Atomic Energy in London. Sir Oliver has since been described by those who knew him as having one of the most brilliant minds of his generation. Before his appointment in May 1948, he represented the United Kingdom at the sixteen-nation conference in Paris in 1947 that determined Europe's response to the Marshall Plan. He served also as chairman of the Organization for European Economic Cooperation, which supervised the division of aid under the Marshall Plan. At forty-three, he was one of the youngest men ever appointed the United Kingdom's ambassador to Washington. He and Acheson would later develop a very close personal relationship. The memo he submitted to Lovett was already in the pipeline before the Joint Chiefs of Staff decision was made. This memo sheds even more light on what was really implied, but not stated, in the Modus Vivendi. The Official Committee protested Carpenter's earlier assertion that if a request were made to extend the nine areas of exchange "the answer would be a definite no." They objected also to Forrestal's observation that when the Modus had been negotiated, stress had been laid on the "general humanitarian aspect of the use of atomic power, rather than on the military side." On the contrary, they argued, the United States had made it clear at the negotiations that its program was a military one, and the British had given full particulars of their own plans to develop two large nuclear piles. Due to the magnitude of the operation, the inference could only have been that the United Kingdom planned to produce plutonium for bomb production. In other words, what was not stated was implied, and "the British Government were under the impression that this had been fully understood by the United States authorities."[32] Hence, the British government was at a loss to explain the ballyhoo surrounding their interest in metallurgy of plutonium and their designs of reactors for the production of plutonium. Moreover, the government could not understand why, after months of useful exchanges of information on matters originally considered to fall under area 8 of the Modus (concerned with the design of natural uranium reactors for the production of plutonium), the AEC should suddenly decide that those matters could no longer be discussed. This development, the memo continued, was even more puzzling when the AEC had initially agreed with Cockcroft's request that Cyril Smith should discuss basic metallurgy of plutonium during his visit to the British installation at Harwell.[33]

Lovett disagreed that exchanges in the area of the basic metallurgy of plutonium were included within the scope of the nine areas. He and members of his staff carefully reviewed the records of the conversations leading up to the Modus and found nothing to indicate that weapons information was envisaged as included in the nine areas.[34] Lovett's observation was correct. Nothing was or would be found. The British argument was based on what was understood or implied during the discussions. It is clear from internal memos that the Americans understood that the British would be engaged in weapons research. But in the face of congressional and military opposition, the State Department, and even the AEC, had to find a way to retreat. Lovett conceded as much. He did not, or could not, explain to Sir Oliver why the exchange of bomb information was initially agreed to by the AEC. He did explain, however, why the administration and the AEC were later forced to interpret the agreement to the letter. He recounted the concerns of the JCAE and the feelings in military circles that a British project would be vulnerable to a Soviet attack. The Joint Chiefs of Staff, he revealed, would be more comfortable if the United Kingdom relocated its project to Canada.[35]

The security concerns of the Joint Chiefs of Staff had become a matter of great sensitivity to the British. In vain the Official Committee on Atomic Energy sought a convincing response to refute or even dilute the vulnerability argument. They considered informing the Americans that if a serious threat of invasion were to develop, all plants would be evacuated and vital components destroyed before they fell into enemy hands. They considered opening up their security arrangements "from top to bottom" to the American authorities. They considered the argument that the threat of destruction in time of warfare was no argument against producing bombs in peacetime.[36] In the end, however, the British conceded that their case was not a strong one. It was impossible for the British Chiefs of Staff to deny the existence of any risk to the security of atomic plants, "and in the absence of any such denial it cannot be easy to convince a doubter that the plants in this country will in fact be secure."[37] Moreover, they agreed that no one had an answer to the American fear that the U.K. government might move further to the left.[38] London decided that it was wiser to avoid answering the security charges unless they were formally asked to do so.

The security concerns of the American Chiefs of Staff were justified. The Labour Party had a long history of leftist sympathies, and, as has been alluded to earlier, some of its members publicly declared that atomic secrets should be shared with the Soviets. Others frowned upon Britain's close ties with the United States, ties that seemed to be getting tighter with each passing year. The Americans could not help seeing a link between British leftists and the Soviet Union. They were concerned because Soviet ambitions threatened Western security. By 1948, not only Bevin but also the British Chiefs of Staff came to accept that the Soviet Union was a threat not only to the security of Eastern Europe but to the security of the United Kingdom and the Middle East. The chiefs agreed that the aim of Kremlin policy was world domination by Soviet-controlled communism. In the event of a war, the Soviets' main objectives would be the conquest of Western Europe, the capture of Middle East oil resources, and the occupation of Scandinavia to protect their flank.[39] The communist coup in Czechoslovakia in early 1948 served to reinforce the suspicions of those who were already distrustful of the Soviets and win over those who had been ambivalent, including a few members of the Labour Party.

The British request for increased collaboration and the objections of the Joint Chiefs of Staff on security grounds were debated in the context of the Berlin crisis. In June, the Soviets placed West Berlin under siege with the objective of expelling the French, British, and Americans from the city. The Berlin crisis served to manifest the often-expressed perception, among the American Joint Chiefs of Staff, of the Soviet Union as an expansionist empire set on European domination. Since Britain stood on the fringe of Europe and was one of the leaders of the Western Union, Britain too was in danger of Soviet conquest. The Joint Chiefs of Staff did not believe that the United States should risk having a British atomic program fall into the hands of the Soviets. The United States, with its stockpile of weapons and its atomic plants in place, had the means to control the Soviets. This was no time to encourage the construction of plutonium plants in the United Kingdom, thereby placing the "secret" within reach of the Soviets.

The United States, however, had no qualms over placing Moscow within reach of American B-29 bombers, modified to transport atomic weapons. In the face of the threat created by the Berlin crisis, the governments of the United States and the United Kingdom agreed to station U.S. bombers in East Anglia in England. (The bombers sent to Britain during the crisis were not yet modified.) Bevin, quickly and uncompromisingly, accepted these bombers, to the surprise of the Americans, who had expected to be presented with conditions safeguarding British national integrity. Marshall even asked Bevin if he had fully considered the implications.[40] Developments in the not too distant future proved that Bevin, caught up in the emotion of the Soviet threat, had not considered the implications. He failed to insist on safeguards to give the British some control over the use of American aircraft and their atomic cargo from British soil in the future. By no means did Bevin single-handedly make the decision. The prime minister, the minister of defense, the Chiefs of Staff, and Herbert Morrison assisted him, but he was the operative force behind the decision. This issue will be revisited in a later chapter.

Nevertheless, the State Department still had not written off collaboration. On November 2, Gordon Arneson recommended to Lovett that steps be taken to improve collaboration. Lovett did not reject outright the recommendation, but he did recognize that the time was not appropriate for a new initiative. Two weeks later, he had an unofficial visit from the British ambassador, who assured him that with or without U.S. help, the United Kingdom would eventually produce a bomb. He did not believe that U.S. officials fully understood that fact. In the wake of new national and international developments, he felt that the whole question of collaboration justified a fresh examination.[41] Among those developments was the just-concluded U.S. elections, which returned Truman to office and gave the Democrats control of both Houses of Congress. The latter development would necessarily result in a change in chairmanship of congressional committees, including the JCAE, which had been headed by Senator Hickenlooper, a Republican. Again, the State Department kept British hopes alive. Lovett did not anticipate an immediate resolution of the problem, especially since the AEC and the military were at loggerheads over custody of the bomb. He suggested that the two governments address the matter in a couple of months, at least after the new congressional committees were organized. Sir Oliver agreed that it was desirable to wait.[42] The decision was an excellent one because the delay led to even more developments that enhanced the chances of closer collaboration. The most important of these developments was Truman's appointment of Dean Acheson to replace Marshall as secretary of state in January 1949.

The year 1948 ended with collaboration in a state of limbo. What had started off as a year full of optimism quickly petered out into more of the same frustrations that by then the British had gotten used to. The disappointments had everything to do with how both sides interpreted the vague and limited terms of the Modus Vivendi. The British wanted more than what was stated in the nine areas because the agreement provided for the inclusion of more at a later date. On the other hand, it seemed that the State Department and the AEC were unable, although at one point they were willing, to honor the provision for an expansion of the nine areas. Moreover, they contradicted their earlier inclination by insisting that weapons information was not covered by the agreement. It is clear that if the "watchdogs" had not intervened, as Groves had done in 1946, collaboration, even in weapons area, would have increased in the second half of 1948. The State Department had to retreat, reappraise its strategy, and plan a new approach to achieve a collaboration that, in 1949, it would seek with more energy than the British.

EIGHT

The United States Proposes Atomic Integration

By the end of 1948, in both London and Washington, government officials came to recognize the failure of the Modus Vivendi. It was clear even to critics of collaboration that the Modus was unsatisfactory. On January 28, Senator Hickenlooper complained to the members of the Atomic Energy Commission (AEC) that the British had benefited from the exchange of information but that he had been "unable to find any area where we have received any substantial benefit."[1] It is quite true that the British had received an enormous amount of scientific material from the AEC. Alec Longair, of the United Kingdom Scientific Mission, sent a memo to James Awberry, of the Ministry of Supply, in which he stated, "I continue to blush at the quantity of material we get from AEC and at the meager return we give them. . . . I do think we should put ourselves out a bit on this and send anything, even if not obviously important."[2] Despite the volume of information received, the British still were totally dissatisfied because the information included no technical material, which they desperately needed for their bomb project. Both sides saw the need for a reappraisal. In November, staff members of the AEC, the State Department, and the Department of Defense all agreed that the whole question of atomic energy should be reexamined.[3]

The timing for a reevaluation was opportune. As discussed earlier, on November 2, 1948, Truman had won reelection by narrowly defeating his challenger Thomas Dewey. Not only did the Republicans fail to capture the White House, but they also lost control of Congress and the chairmanship of its various committees. In January, Senator Brien McMahon, a Democrat, replaced Hickenlooper as chairman of the Joint Congressional Committee for Atomic Energy (JCAE). The changes did not end with the results of the election. William Webster replaced Donald Carpenter as chairman of the Military Liaison Committee and immediately made his presence felt by proposing an atomic energy meeting at Princeton, New Jersey, where the State and Defense Departments and the AEC could air their views on collaboration.

The State Department itself experienced a major shake-up. Acheson replaced the ailing Marshall as secretary of state, and James E. Webb succeeded Undersecretary of State Lovett. Acheson had always been sympathetic to the British cause. He never did reconcile himself to the U.S. failure to abide by the terms of the Quebec Agreement. He would later state that his knowledge of the Quebec Agreement would disturb him for years to come, "for with knowledge came the belief that our Government, having made an agreement from which it had gained immeasurably, was not keeping its word and performing its obligations." He admitted that "grave consequences might follow upon keeping our word, but the idea of not keeping it was repulsive to me."[4] By March, illness would force Forrestal too to resign. Louis A. Johnson replaced him as secretary of defense. For a while Johnson fell in line behind the State Department's and the AEC's atomic policy until, gradually, his Anglophobic sentiments revealed themselves.

The new year had barely started when steps were taken to address the collaboration issue. On Webster's recommendation, the first of two conferences was convened. These meetings are very significant because their recommendations led to a new policy for improved atomic relations with the British and Canadians that the president, the State Department, the AEC, and, ultimately, even the Defense Department would embrace. Representatives from these departments met at Princeton on January 24 and 25. The members included Robert Oppenheimer, chairman of the AEC's General Advisory Committee (GAC); James Conant, still president of Harvard University and a member of the GAC; from the State Department, Kennan, Arneson, and George Butler, deputy director of the Policy Planning Staff; from the AEC, Carroll Wilson and Joseph Volpe; and from Defense, General Lauris Norstad, General Kenneth Nichols, and Webster. They already had in hand a working paper that drew their attention to U.S. uranium needs. If the United Kingdom got involved in the production of atomic weapons, the paper pointed out, the result would be a drain on joint uranium supplies. Moreover, present allocation arrangements, as provided for in the Modus, were due to expire at the end of 1949, when a new agreement would have to be worked out.[5]

The ore question was of critical importance to the United States. So far it had not become necessary for the United States to draw on the British stockpile. However, long overdue developmental work was due to begin on the Congo mines in January, and a considerable reduction in the Congo output was anticipated. Arneson felt certain that the United States would have to call on U.K. stocks to an amount between 600 to 1,000 tons to make up the Congo shortfall.[6] Thus, it was in the U.S. interest to improve relations in order to pave the way for a smooth transfer of these stocks, not only in 1949 but also in 1950, after the expiration of the Modus. The U.S. officials did not view the shortage as an insurmountable problem. They estimated that on the basis of the needs of the U.S. and U.K. atomic programs, there should be sufficient production from the Belgian Congo, Canada, the United States, and South Africa to supply both programs for the period 1949 through 1955. But this optimistic estimation was based on the success of the Redox system—a plan for the recycling of depleted uranium. If Redox failed, they conceded, there would be insufficient supplies to go around. Only the U.S. program would be sustainable.[7]

The Princeton Conference surmised, among other things, that atomic energy plants in the United Kingdom would be vulnerable in the event of a war, that U.S. plants were more efficient in the conversion of raw materials, and that duplicating American plants in the United Kingdom would be an uneconomic diversion of British technical and economic resources. On the other hand, the report conceded, it would be futile to attempt to stop the U.K. program by exerting U.S. pressure. The consensus was that the limited cooperation provided by the terms of the Modus should be discontinued. In its place, the report proposed a complete interchange of information in all fields of atomic energy, including weapons. All production facilities were to be located with due regard for strategic considerations. And the atomic energy programs of both nations were to be coordinated so as to make the most effective use of joint resources, specifically raw materials. In addition, the report recognized that it would be "dangerous and unsound" to attempt any revisions of the existing arrangements without "full public disclosure of the relevant facts." Moreover, congressional blessing had to be sought to confirm any new action taken.[8]

A week later, the full AEC endorsed the recommendations. It authorized Lilienthal to urge the secretary of state to initiate discussions as soon as feasible with the United Kingdom and Canada.[9] The lone dissenter was Strauss, who submitted his own memorandum to the other members of the AEC. He argued that to avoid misunderstandings and disappointments in the long term, the British should first agree that atomic weapons would not be manufactured in Britain and that fissionable material would be neither manufactured nor stockpiled in Britain. Only with those guarantees in hand should the United States agree to enter into a partnership.[10]

On the surface, it appeared that Strauss was not the only one who harbored reservations. Truman summoned Lilienthal on February 9, ostensibly to discuss the appointment of two new members to the AEC. Lilienthal reported that he was "caught off balance" when the president began to discuss an entirely different subject. Truman reminded him that the atomic bomb was probably the only instrument preventing the Russians from overrunning Europe and that hence he had to guard it very carefully. He disclosed that Senators Millard Tydings and Connally had criticized the AEC and Lilienthal for releasing too much atomic-related information in their public reports. The president then drew Lilienthal's attention to the Modus and "went on to say that we have got to protect our information and we certainly must try to see that the British do not have information with which to build those atomic weapons in England because they might be captured."[11] Was Truman, in February 1949, still adamantly against collaboration, or was he attempting to force Lilienthal to convince him of the merit of cooperation? His conversation with Lilienthal is somewhat puzzling, since subsequent developments in the immediate and distant future tended to contradict his opposition to cooperation.

Acheson had already started the process of convincing him of the security benefits of reevaluating his policy. And certainly he had struck the right chords with the president. On the day following the Lilienthal discussion, Truman approved the appointment of a Special Committee of the National Security Council to prepare recommendations to improve atomic relations among the United States, the United Kingdom, and Canada. Acheson suggested, and he agreed, that the committee should consist of the secretary of state, the secretary of defense, and the chairman of the AEC—the American members of the Combined Policy Committee (CPC). In turn, each departmental head appointed two staff members from his respective department to help explore the matter. The working group established by this procedure consisted of Kennan and Arneson (State), William Webster and Kenneth Nichols (Defense), and Carroll Wilson and Joseph Volpe (AEC). General Eisenhower, presiding officer of the Joint Chiefs of Staff, was also invited.

Within three weeks the Special Committee submitted its report, which was accepted by the State Department, the Defense Department, and the AEC. On March 2, Sidney Souers, executive secretary of the committee, sent it on to Truman, who endorsed it as the official policy of his administration with regard to atomic cooperation with Britain and Canada. The report repeated most of the recommendations of the Princeton Conference and added a few of its own. It justified its recommendation for atomic collaboration on the grounds that extensive international cooperation had existed, did exist, and would continue to exist between the United States, the United Kingdom, and Canada. If collective security arrangements were to be pursued by the three nations, it would not be in keeping with a "liberal interpretation of the spirit of confidence and good faith" if the United States refused to collaborate fully with the United Kingdom and Canada on atomic weapons. Moreover, since joint strategic plans were being worked out on the assumption that atomic weapons would play a vital role, and since the United Kingdom was about to develop atomic weapons of its own, it would be illogical to exclude atomic weapons' production from the close collaboration among the three countries.[12]

The report repeated the U.S. objections to the existence of a U.K. weapons program: vulnerability to enemy attack, inefficient use of scarce raw material, and inefficient diversion of U.K. technical and economic resources. But it concluded that termination of collaboration could "have the most serious consequences in terms of jeopardizing the availability of raw materials." The raw material needs of both countries through 1951 would be sufficient only if neither country expanded its present program and if the United States continued to have access to the U.K. stockpile in excess of the present requirement of the U.K. program. Eventually, the United Kingdom would produce its own atomic bomb; however, the report cautioned, without U.S. help, the United Kingdom would retain complete freedom of action and might "indeed become a serious competitor with us for scarce raw materials."[13]

The Special Committee suggested to the president the establishment of full cooperation in all fields of atomic energy under the terms outlined by the Princeton Conference. Those terms included (1) locating all fissionable materials, production plants, large-scale atomic energy developments, and supplies of strategic material in either the United States or Canada; (2) locating any expanded production program in either the United States or Canada; and (3) storing all nuclear components of atomic weapons in either the United States or Canada. The report calculated that over the next five years the U.K.-Canadian effort would require no more than 10 percent of available raw material. Meanwhile, the United States would use the other 90 percent in the major effort of producing atomic weapons for joint defense.[14] In effect, the plan called for the United Kingdom to integrate its program into that of the United States, a return to the wartime arrangement that had caused immeasurable frustrations. The proposal envisaged the United Kingdom as a sort of American atomic colony whose raw materials and technical experts would be exploited to feed the atomic plants of the United States and to manufacture U.S. atomic bombs.

It was still unknown to the Americans how the British would react to this new initiative; their advice had not been solicited in the formulation of this new American policy. We know that even if the United States had chosen to keep the Modus in place, the British would have rejected that option. The United Kingdom had already asked officially for an extension of the nine areas of information exchanges, an extension that they believed was provided for in the terms of the agreement. The United States had not yet responded officially to that request; rather, the United Kingdom had been asked to wait until the election had effected its changes. The British waited patiently, knowing that if nothing was done before the expiration of the Modus, the existing arrangements would revert to the 50–50 plan of the pre-Modus days. At least, they would be well provided for with uranium, an assurance they knew the Americans did not have.[15] They knew also that the Americans were working on a resolution and that a formal invitation to engage in talks was imminent. On March 9, Acheson had confirmed to Sir Oliver Franks, the British ambassador, that atomic energy talks were in the works.[16] Later that day, Arneson presented Sir Gordon Munro with a detailed account of the stages that the United States proposed to follow before sitting down with the British. Congress had to be informed first, albeit in general terms. The administration had to devise a plan to win over the congressional leaders, without whose blessing any new arrangement was worthless. Arneson assured Sir Gordon that the British would not be presented with a "take it or leave it" proposition.[17]

In early April, the subject was raised again, this time by the Americans, as an initial step in selling the new policy to the British. In the course of a visit to the United Kingdom, General Lauris Norstad, with the blessing of Acheson, Forrestal, Eisenhower, and Webster, discussed with Marshall of the Royal Air Force Lord Tedder, British Chief of Air Staff, the question of the location of large-scale atomic energy facilities. Norstad intimated that American military authorities were troubled by the United Kingdom's plans to develop a rather major atomic energy and atomic weapons program. Revealingly, in terms of insight into British consensus, Tedder replied that he too was troubled and that he would pass on the American concerns to Tizard, who had an important voice in British atomic activities. Webster was optimistic that Norstad's talk with Tedder would influence the British to adopt a position closer to that of the United States once negotiations began.[18] There were good reasons for optimism, since, according to Norstad's official report to Acheson, Tedder's views seemed to accord with those of the Special Committee. Tedder was of the opinion that if a fully effective atomic energy partnership were introduced the British would probably not insist on having a major production program. From a military standpoint, Tedder did not believe that it was wise for the British to store major quantities of bombs in Britain. He did agree with an observation of the Special Committee that, to command public support and bolster national prestige, some plutonium production work would have to be undertaken in the United Kingdom. Norstad urged Tedder to recommend to his people the formulation of a minimum position rather than a major negotiable demand and stated that he would do the same on the U.S. side. Acheson was pleased with the report.[19]

In mid-April Truman approved the Special Committee's report, but he did not submit it to Congress for three more months. The administration believed the climate was not right, since Congress was about to debate the North Atlantic Treaty agreement. The following month the congressional climate became even less tolerable. Senator Hickenlooper, still very vigilant in his role as atomic watchdog, had leveled the charge of "incredible mismanagement" against the AEC. Among other things, the commission was accused of giving away too much information to the British and of being unable to account for the loss of a bottle of uranium oxide from the Argonne Laboratory. Lilienthal publicly demanded a congressional investigation to clear the AEC. The investigation dragged on for about four months, during which "the public and even senators on the investigative commission lost interest. . . . At some sessions only three [of the fifteen] members attended."[20] The investigation turned up no "incredible mismanagement." But it did slow the momentum of the new Anglo-American initiative. It was evident even to the British that cautionary steps were needed.[21]

The administration was ready by July. Acting on Acheson's suggestion, it attempted to bypass the full JCAE and the full Congress by discussing the new policy with congressional leaders from both Houses and parties. McMahon, chairman of the JCAE and one who tended to support the president's atomic policy, felt that the JCAE would not be receptive and thought the idea of bypassing it an excellent one.[22] This meeting convened at Blair House[23] on the evening of July 14.[24] The president opened the meeting by stating his full support for the proposed solution. During the next two hours, the representatives of the executive branch hammered home two main points: the raw material shortage and the security issue. Lilienthal focused on the importance of the United Kingdom as a guarantor of U.S. raw material, while Eisenhower stated his conviction that in the event of another global war the United States would be lost without the British. He could not understand how the United States could link its military fate with the United Kingdom and still exclude atomic weapons from their full partnership. Senator Vandenberg, weakly supported by McMahon and Tydings, wondered aloud whether the British would accept an alternative arrangement whereby the United States would continue to be the sole producer of atomic weapons in exchange for earmarking a certain number for British use. This suggestion was destined to gain adherents in the future, but at this meeting Acheson shot it down as unrealistic. He did not believe that the British would accept a proposal that would mean the downfall of their government: "No Government which had so near at hand such an important weapon for the defense of the British Isles would voluntarily surrender this development into the hands of another country, however friendly."

Uncharacteristically, Senator Hickenlooper said nothing until late in the discussions, and only after he was prodded by Vandenberg. He stated emphatically that he was against the proposal and that it was contrary to the Atomic Energy Act of 1946. He did not think that the raw material shortages were serious enough "to warrant giving away to the British our greatest heritage and asset." Lilienthal took strong issue with that charge. He explained that if the Modus ended, the U.S. atomic program would slow down within three months and large numbers of workers would have to be laid off. General Eisenhower interjected: "And who would take responsibility for explaining 'that' to the American people?"[25] McMahon also shared Hickenlooper's skepticism concerning the seriousness of the raw material problem. In a letter to the secretary of defense, dated that same day, in which McMahon strongly endorsed an increased production of atomic weapons, he too raised his doubts. He stated, "It will of course be said that we lack adequate raw materials to implement a program of this kind. But the argument does not impress me. Recent progress in developing methods of extracting uranium from phosphate shales, plus other analogous advances, plus the availability of ores in South Africa and other friendly foreign countries, indicates that if we are willing to expend the necessary effort, we can break the raw materials bottleneck."[26] Despite his own concerns, McMahon did not speak out in support of Hickenlooper. In the end, the majority agreed that negotiations should be entered into with the British and Canadians, but only after the matter was presented first to the full JCAE. That group could not be bypassed.

As the meeting broke up, the president swore the participants to secrecy, but already there were about 50 newspaper reporters on the sidewalk "creating an air of excitement." Lilienthal, for one, was convinced that the particulars of the meeting would be leaked to the press.[27] He was right. The State Department suspected that Hickenlooper was the culprit.[28] The immediate fallout was the threat of resignation from the JCAE by Senators William Knowland of California and Eugene Millikin of Colorado. Although McMahon regarded their letters of resignation as a "stageshow,"[29] the administration recognized the critical need to consult the full JCAE before McMahon lost control of its members. A meeting was arranged for the 20th of the following week.

The administration was equally concerned about foreign reactions to the leak. It feared that France might become "disheveled" over its exclusion from a tripartite meeting. The French were already taking steps to organize their own atomic club of smaller Western European nations—a move that the

United States opposed on raw material grounds. However, a more alarming result of the leak was felt in Belgium. The Belgian press and opposition again raised the sensitive question of the sale of Belgian ores to the United States and United Kingdom. The new Belgian coalition government under Prime Minister Gaston Eyskens denied the existence of any secret treaty with foreign countries. Eyskens's denial raised questions in the U.S. administration as to the extent of his knowledge of the relevant agreements.[30] The administration feared that the outgoing government had not briefed him. Nevertheless, the United States did not view his lack of knowledge as a threat to the raw material agreement, since it was not due to expire until 1956. The time frame gave the United States sufficient time to develop new sources of uranium.

The long-awaited meeting with the full JCAE took place at Blair House on the afternoon of July 20.[31] The meeting turned out to be memorable in terms of its role in influencing the administration's atomic policy and in terms of the issues it raised concerning the power of the president to shape foreign policy, as it related to atomic energy, without the consent of Congress. It was a contentious affair, one that the executive, by its own admission, lost. From the outset, the legislative members seized the initiative from Acheson, totally disrupting his well-planned agenda. It was clear that the question that most concerned the congressmen was the legality of any atomic energy overtures to the British. Senator Knowland interrupted Acheson's opening statement to ask whether in the opinion of the State Department the Atomic Energy Act of 1946 permitted the proposed arrangements. Acheson was able to dodge a detailed answer. When Lilienthal addressed the raw material needs of the United States, Knowland again broached the subject by asking whether the proposed course of action could be taken without formal action by Congress. Lilienthal replied that the AEC was guided by the findings of the executive branch that the proposed arrangements were consistent and contributed to the common defense and security of the United States.

Later, when Acheson attempted to read the conclusions contained in the Special Committee's report, he was again interrupted by questions from several of the congressmen. McMahon was curious about the feasibility of dismantling the U.K. plants and duplicating them in Canada. Hickenlooper did not believe that the United States should have to bargain with the United Kingdom for uranium that belonged to the Belgians. Congressman Cole inquired whether both parties had to consult each other before using the bomb. Congressman Carl Hinshaw wished to know how much public information was disseminated on the whole collaboration issue. Acheson never did finish his presentation. Knowland threatened to oppose any negotiations unless the State Department gave advance commitment that it would secure congressional approval for any arrangement that might evolve from the negotiations. Senator Millikin interjected that he would raise the matter in Congress if the State Department went ahead without those assurances.

Acheson could evade the question no longer. He explained that the president did have the legal right to hold discussions with the British because the proposed arrangement was calculated to promote the common defense and security, a condition that overrode the strictures of the McMahon Act. He stated that the permissibility question had been raised earlier in connection with the Modus and that the JCAE had agreed then that the Modus arrangements were permitted under the law. Millikin insisted that Congress had preempted the field of atomic energy and hence that negotiations could not be started without the consent of Congress. Congressman Elson, for one, "took violent exception" to Acheson's interpretation of the law. Lilienthal would later report that when Acheson presented the executive's view of the legal question, "Knowland blew his top—the redness in his face changed to pale white." He threatened to go to Congress—"yelling and pretty out of control."[32]

By then it was clear to the proponents of the proposed negotiations that a consensus was far out of reach. Even within the ranks of the AEC, dissension arose. Commissioner Strauss challenged Lilienthal's statement that the AEC supported the new objective. Strauss explained that while he had been a minority of one in the past, that position might change with the addition of the two new members to the AEC.[33] Gordon Dean and Henry Smyth, the new commissioners, said nothing to dispute Strauss's observation. Louis Johnson, the secretary of defense, separated himself from Acheson by suggesting to the JCAE that the defense establishment would review its entire position in view of congressional opposition.[34] This meeting marked Johnson's first tentative step in the direction of opposing the initiative. Acheson and his State Department team viewed his retreat as "perhaps the most serious development."[35] The consensus at the State Department was that before the full JCAE, Eisenhower was not as effective as he had been at the Blair House meeting, where he had stressed the importance of the Anglo-American alliance and other broader implications. He tended to reduce the issue to one of atomic energy information in exchange for uranium; or, rather, he fell into the trap of the JCAE's interpretation of the whole issue. Acheson surmised that Eisenhower had "retreated into a simple soldier unacquainted with complexities."[36] Acheson himself classified the meeting as "a failure just short of a disaster." He admitted that "our forces were shaken, and I managed our side badly when the engagement became disorderly."[37]

The administration should have anticipated intense opposition from the

JCAE, coming as the meeting did in the waning weeks of the congressional investigation of the AEC. By early July, it was clear that Hickenlooper would not be able to prove his charges of "gross mismanagement." The media had begun to lose interest, and in desperation Hickenlooper had resorted to fishing for evidence in unlikely places, such as "house-keeping shortcomings at AEC installations, including garbage cans."[38] Moreover, the investigation had clearly shown that the JCAE was still very much committed to its appointed task of guarding the "secret." Hence, to expect the congressional members to jump on the cooperation wagon was somewhat optimistic. Hickenlooper, for example, with defeat at his instigated hearings staring him in the face, must have welcomed the meeting as a means to redeem his credibility and revive his battered leadership among his congressional colleagues.

The State Department knew where Congress stood on sharing atomic energy with the British. The issues raised by the JCAE reinforced long-held opinions. Its members had always viewed and described the British as economic supplicants. Time and time again they had implored the State Department to use its economic weapons to force the British in line. Moreover, senators tended to approach the entire atomic energy field solely from the perspective of U.S. national security. The JCAE's members, unlike the State Department's, never weighed the national prestige issues from the United Kingdom's perspective. Indeed, theirs was a narrow, and highly nationalistic, interpretation of Anglo-American atomic relations. Thus, one must conclude that although a meeting with congressional leaders was necessary, it was ill timed and mismanaged.

Senator McMahon, an ally of the executive branch, had his doubts about the proposal. He found himself in a delicate position at the meeting. On the one hand, he tried to avoid abrogating his responsibilities as chairman of the JCAE; on the other, he showed sensitivity to the administration's goals. But it was clear he did not believe that the British should be in the business of producing atomic bombs. Three weeks later, during an executive session of the Senate Armed Services Committee, the Senate Foreign Relations Committee, and the Joint Chiefs of Staff to consider the Military Assistance Program and the North Atlantic Pact, he stated his view, and Vandenberg agreed. The British should be forced to agree to leave bomb production to the United States before the United States agreed to give them any kind of military assistance.[39] The State Department was forced to issue a policy statement shooting down the idea before it gained any momentum.[40]

Over the next few days the administration's carefully thought-out atomic energy plan quickly unraveled. The administration was not willing to force a confrontation over the issue of legislative and executive jurisdiction. Despite controlling both Houses and despite the president's willingness to accept the challenge to his authority, the State Department was against contesting the legal question.[41] It decided instead to modify the proposal to win over the JCAE members. Lilienthal went along reluctantly. He was not overly optimistic that the JCAE would readily accept the new proposal, "as having tasted blood they may want to move in and have the kill."[42] Lilienthal was in favor of the president's taking the issue to the general public, where, he believed, sentiment was ready to support a collaboration effort.[43] Secretary Johnson too accepted the modified proposal, and on July 25 he and Acheson presented it to a skeptical president.

They advised Truman not to press the point of his legal rights. Instead, he should focus on winning over the JCAE with the explanation that the administration had never intended to act unilaterally but rather in a bipartisan manner. Exploratory and frank talks would be conducted with the United Kingdom to ascertain how the critical uranium needs of the United States could be met. The administration would then refer the result of the exploratory talks back to the JCAE before any firm agreements were made. Meanwhile, the British and Canadians would be asked to agree to a continuation of the Modus for some months after its expiration at the end of the year.[44] On July 27, the full JCAE accepted the new proposal. Knowland tried to get an even stronger statement but found no support among his colleagues.[45]

It was not easy getting the president to accept the compromise. He believed that he had a constitutional and legal right to pursue the original policy. He believed that the common defense and security section of the McMahon Act gave him the right to open negotiations with the British and Canadians. Hence, when the legislators were called in, it was not to get their permission to begin the talks. Rather, it was to inform them that the executive branch intended to act on the recommendation of the Special Committee and to explain the basis for accepting the recommendation. As late as the day on which Acheson was due to present the compromise to the JCAE, he was on the phone pleading with Truman to agree to it. It is clear that the president backed down only because he was entreated to do so by Johnson and Acheson. A few weeks after the JCAE accepted the compromise, Acheson attempted to get the president to review the entire policy that had come out of the Special Committee. The president refused on the grounds that the policy was correct. He told Acheson that while the talks with the British were proceeding, steps should be taken also to inform public opinion. After all the facts were in, he would review the situation in Congress and determine the future procedural course of the United States.[46] He was still not fully committed to congressional involvement.

On the other hand, his secretary of defense was not convinced that the United States should share atomic secrets with the British. He did not make this opposition explicitly clear until several weeks after the first JCAE meeting. Meanwhile, he left State Department officials in a state of uncertainty concerning his stand on the proposed arrangements. Military personnel were called in, but no one seemed to know his view.[47] Finally, on August 12, Undersecretary of State Webb, with the blessing of his colleagues, decided to confront him. At a luncheon attended by Johnson, Admiral Sidney Souers of the National Security Council, Stephen Early, deputy secretary of defense, and General James Burns, assistant to the secretary of defense, Johnson gave vent to his feelings. In an outburst he stated flatly that the U.S. position in the tripartite talks was wrong and that he doubted that he would go along with it. The United Kingdom was finished, and he saw no sense in trying to bolster her up with assistance programs, including atomic energy assistance.[48] At that same luncheon, according to Lilienthal, Johnson also added that there was no legal authority for the Modus and that the AEC was eager to "give away the secret."[49] All of this indicates that the secretary of defense had gone along with the new arrangement only because the president and other members of the administration did. His feelings toward the British were deep-seated and not just the result of an impulsive outburst. It partly explains also why he was so quick to compromise at the JCAE meeting and why he unhesitatingly supported Acheson in his representation to the president to grant the JCAE veto powers over any new arrangement.

The proposals presented to the British[50] at the full CPC meeting of September 20 departed very little from the recommendations of the Special Committee's report. The Americans asked the British to agree to an extension of the expiration date of the Modus until such time as it could be replaced by a new long-term agreement. Sir Oliver's response set the tone for British recalcitrance at future sessions. The extension requested by Webb would be considered only after the exploratory talks had reviewed the whole field of collaboration.[51] Over the next ten days, the British negotiators made their position perfectly clear. They were willing to recommend to London that the government drop a recently approved third pile from its program and focus instead on two piles plus a low-separation plant. Such a reduction would cut down on the use of raw material by the United Kingdom and so increase the supply for the U.S. expanded program. However, they were unwilling to consider transferring their weapon production facilities to Canada or relinquishing the production of atomic weapons to the United States. The British government had already decided, and had informed the United States of its decision, to have an independent atomic energy program in the United Kingdom. Also rejected was the American insistence that all atomic weapons and components should be stored in the Western Hemisphere. Sir Oliver and Sir Roger Makins made it quite clear to Kennan that the British had every intention of storing atomic weapons in the United Kingdom. Finally, they demanded full and complete cooperation, without any restrictions, and including the design, production, storage, and delivery of atomic weapons. Moreover, Sir Roger implied diplomatically, if the countries could not come to an agreement, the British might be much more cautious over uranium allocations.[52]

The British position left the American negotiators under a dark cloud of pessimism, dashing all hopes for a quick and amicable agreement. The Americans were not impressed by the British magnanimous gesture of offering to drop their third pile. They felt that the British were offering "the abandonment of practically an imaginary plant" that had not proceeded beyond the theoretical stage and that they in return expected to receive a full exchange of information. Moreover, even if the British proceeded with a third pile, it would not require supplies of raw material before 1953. Furthermore, they saw no need for the British to continue receiving 50 percent of allocation, their right under the Modus, because their two piles would not require additional supplies in excess of what could be provided for from the British stockpile. The suspicion was that the British were trying to apply pressure on the United States in its time of need.[53] The CPC adjourned on September 30, affording its members an opportunity to report to their respective governments.

The president did not view the United Kingdom's demand with the same degree of pessimism as his officials. He asked Kennan and Webb to sound out the JCAE to determine whether its members would endorse the British position. But he was still not too happy about having to win the approval of the JCAE for his foreign policy. At a cabinet luncheon later that day, he revisited the legislative-executive issue. He reiterated that he had the legal authority to conclude an agreement without the consent of the legislators. He thought that the provision of the law restricting or limiting his powers was unconstitutional. Attorney General J. Howard McGrath and Vice President Barkley cautioned him against proceeding without the JCAE's consultation and approval.[54] A week later, Truman again repeated his determination to follow an atomic energy policy that he and his advisers, and not the JCAE, deemed to be in the national interest of the United States. He told Webb that he had spent the weekend on the presidential yacht, the *Williamsburg,* where he had given careful thought to the problem. He felt that it was essential to work out a "thorough-going partnership" with the United Kingdom and that, if necessary, he would attempt to get the backing of the country to override congressional opposition.[55]

Despite the president's optimism, garnering public support might not have been as easy as he, or even Lilienthal, anticipated. A Gallup poll in early September already had shown that 72 percent of those polled overwhelmingly disapproved of the idea of sharing atomic secrets with the British. The same poll revealed that 81 percent were against the shipping of atomic weapons to nations that were friendly to the United States.[56] It is uncertain whether Truman had seen these results. If he had, then he could have been gambling that public opinion had changed significantly in the face of startling new developments in the Soviet Union.

By September 19, the eve of the exploratory talks, the British and Americans had confirmed that the Soviets had tested an atomic bomb in late August. At the risk of an understatement, the discovery took both administrations by surprise. The U.S. monopoly of atomic weapons came to an end sooner than the American experts had predicted. Truman admitted that though the intelligence experts had different opinions, none of them had anticipated that the Russians would detonate any atomic device before 1952.[57] As a matter of fact, a Central Intelligence Agency report, dated July 1, 1949, projected that the Soviets' first atomic bomb could not be produced before mid-1951.[58] One year earlier, Forrestal had had a conversation with Bedell Smith, the U.S. ambassador to Russia, who had assured him that the Russians did not have the "industrial competence" to produce an atomic bomb. "They may well now have the 'notebook' know-how, but not the industrial complex to translate that abstract knowledge into concrete weapons."[59] Professor Niels Bohr, the Danish physicist who had acted as an advisor to the Manhattan Project, was one of those who did not underestimate the scientific capability of the Soviets. In October 1947, he advised John McCloy, former assistant secretary of war, that the West should not overlook the enormous concentration of effort that the Russian totalitarian system could bring to bear upon its project. He believed that the Russians would develop a bomb within sixteen to eighteen months.[60] He was not off by more than four months.

What effect did the Soviet explosion have on the guardians of the secret? The explosion of the Russian bomb, Acheson later wrote, also exploded "a good deal of senatorial nonsense about our priceless secret heritage."[61] But did the explosion engender among the members of the JCAE a willingness for greater cooperation? Arneson seemed to think that it did. In a memorandum to Webb, he stated that it had had a "sobering effect on members of the JCAE. Their reactions to a partnership arrangement with the United Kingdom and Canada appear more hopeful than was the case some weeks ago."[62] When Acheson made a full report to its members on the status of the talks with the British and Canadians, he too found a rather subdued gathering. They did not appear to be taken aback by the British position. Rather, they voiced satisfaction that the discussions had been conducted in complete good faith and in accordance with the established guidelines. The consensus was that further exploratory discussions should continue with the ultimate goal of persuading the British to accept an American monopoly of bomb production.[63] The president was pleased when Acheson gave him the result of the talks.[64] A few days earlier, when the AEC reported to the JCAE on the talks' progress, it too had received an endorsement. Hickenlooper even suggested that the British be given the assurance that if weapons were made in the United States they would be made available to the United Kingdom in time of need.[65] He did not mention who would actually control the weapons. It should be emphasized, however, that the Soviet detonation had not forced a reappraisal in terms of the location of production facilities. If anything, the security of the British Isles had been compromised even further, since the island now had become susceptible not just to conventional Soviet bombers but also to atomic attacks. Hence, producing bombs in the United Kingdom seemed even more impractical to the JCAE. All the same, the JCAE clearly was coming to grips with the realization that it no longer had a "secret" to protect. Either that, or its members were still in a state of shock. Vandenberg had already confessed to Ernest Gross that the effect upon the committee of the news of the explosion was "paralysing."[66]

Across the Atlantic, the British government made a futile attempt to persuade Truman to forgo an official announcement of the explosion and instead opt for a leak to the media. On the eve of the revelation, Sir Roger flew to New York from Washington to argue with Acheson that the former approach might generate alarm and panic among the American public while the latter would lead to a more gradual realization that the secret was out.[67] In fact, Sir Roger was more concerned about public reaction in Britain. The British government was in the midst of its own public crisis surrounding its devaluation of the pound. The government feared that a sudden disclosure would lead to even further furor over the shortcomings of a Labour government that had not yet produced a bomb. But Truman had already made up his mind. A joint declaration was made on September 23, in which both governments again emphasized the need for international control.

The Soviet detonation left the British even more astonished than the Americans. At least the Americans could reconcile themselves to the knowledge that they possessed an arsenal of atomic weapons. The British, on the other hand, were still struggling to master the technical intricacies of a program that they had thought to be years ahead of the Soviets'. Sir Henry Tizard was convinced that the Soviets had acquired full details of the American process. The only plausible explanation for the Russian achievement, he believed, was that they had managed "to steal plutonium from the United States. From what I have heard unofficially I should think that it was quite possible."[68] British incredulity is quite understandable. The British too had accepted the American projection for the completion of a Soviet bomb. After all, the Soviets faced the formidable task of having to start from scratch; they had not been a part of the Los Alamos team. The only way they could have jumped ahead of the British, according to Tizard's hypothesis, was by spying and not by more advanced technological expertise. Within four months the truth of his statement surfaced. All the same, the Soviet achievement forced the British to reevaluate their atomic program in terms of its immediate role in British defense policy. As things stood, their island was virtually defenseless against a Soviet attack unless their American ally honored the terms of the untested North Atlantic Treaty and rushed to their defense in an emergency. Even with an American commitment, the British still were vulnerable until American atomic weapons actually were shipped to Britain. It soon became clear to British policy makers that atomic weapons were desperately needed on British soil to restore a European balance of power. Unfortunately, the best British estimate did not project the completion of a bomb before 1952. Some deep soul-searching was in order, and time was of the essence.

Both the Americans and the British did some preparatory work before scheduling the next round of the exploratory talks. The prime minister ordered a study into the effect on the British atomic program of dropping the third pile. The result was reassuring. The government would save an enormous sum of money. Between 1950 and 1956 a capital outlay of 11 million pounds ($22 million) would be saved, plus a savings in annual operating costs of between 250,000 pounds ($.5 million) in 1950–51, rising to a high of 2,500,000 ($5 million) in 1954–55. Moreover, any loss in processed uranium would be made up from the receipt of enriched uranium 235 from the United States that might flow from a program of full cooperation. In addition, access to American technical information would enable the United Kingdom to design and manufacture weapons with much greater striking force than would otherwise be possible.[69]

For their part, the Americans, on "invitation" from the British government, sent over three of their people to inspect the British facilities on November 7 through 10. The Americans used the visit to determine the extent to which the British would be able to contribute to the U.S. program. It was the sort of ammunition the administration needed to buttress its case before the JCAE. The British were not too happy about opening up their facilities to the Americans without the assurance of an eventual agreement, but the risk had to be taken to prove that their piles were of a high technical level.[70] The inspectors' report to Washington was highly positive. They reported that the two British plutonium production piles were too far advanced to recommend termination and, inasmuch as the British method was different, the United States stood to gain new information and plutonium by continuing the two piles. The low-separation diffusion plant, on the other hand, could be canceled without appreciable loss. Furthermore, "in the field of fuzing, ballistics, detonators, and possible explosives production," work in the United Kingdom could be of considerable value to a combined effort. They recommended that research work at Harwell continue because there were "ample opportunities for gain from a many-sided approach in research."[71] Thus, the United States stood to gain from full collaboration with the United Kingdom.

The exploratory talks resumed on November 28. Two main problems were addressed: the question of an interim allocation of ore to the United States and the issue of long-term cooperation. The allocation problem did present a major problem, since the existing agreement was due to expire in four weeks. Moreover, the administration was actively engaged in a new debate concerning whether the United States should develop a super- or hydrogen bomb. Its proponents argued that the United States needed a hydrogen bomb to keep it ahead of the Soviets, who had caught up in atomic technology. Until the superbomb was developed, the United States could maintain its superiority primarily by increasing the number of bombs in its arsenal.

The raw materials issue became somewhat acrimonious. The Americans placed the British on the defensive by accusing them of using raw materials to pressure the United States. Certain unnamed senators had already leaked to the press their fears that the British might use raw materials as a "bargaining point" in the negotiations. Earlier, both Britain and Canada had been forced to reassure the United States that no such plan was afoot.[72] At the talks, the British again denied the charge, which did have a measure of truth to it. The British plan had been to hold out on raw material allocation until the long-term issue became clearer.[73] They were fully aware of the importance of an increased supply of uranium to the U.S. expanded program. They were also aware that they could not create a negative atmosphere by allowing the perception of blackmail to pervade the talks. The U.K. delegates eventually agreed to the appointment of a raw materials subgroup to look into the full range of the allocation issue. By the end of September, they agreed to allow the United States to keep all the Congo ore produced in 1950. This decision was taken only after Acheson personally intervened to ask Sir Oliver to expedite the issue before it festered to the point of embittering congressional opinion.[74]

With regard to the American proposal of September for renewed cooperation, the British repeated their earlier position. They preferred the establishment of two mutually supporting programs, one in the United States and the other in the United Kingdom. They envisaged the establishment of a full atomic bomb program within the United Kingdom, a program that would be able "to complete the whole process from ore to weapons."[75] That was not the answer the Americans were looking for. As far as they were concerned, any such program would be a wasteful duplication of effort.[76] Finally, the United States presented its own counterproposal. It emphasized its preference for integrating both programs along the lines of what had existed during the war. The British delegates were all ears and full of questions.[77] Over the next few days, the two delegations fine-tuned the new proposal into a full-scale draft for presentation to the British government. Its main features included the following:

1. The objective of the integration program was to produce the maximum number of weapons in the shortest possible time.
2. New lines of atomic bomb development had been opened up, but the United States lacked an adequate number of first-class scientists to explore all the areas.
3. The United States therefore proposed that a number of first-class U.K. scientists, and their staff, should be seconded to work at Los Alamos and Sandia.
4. The United Kingdom would continue to develop that part of its atomic weapons program that contributed to the common interest. However, the United Kingdom would not duplicate work already carried out in the United States.
5. The United Kingdom would not be prohibited from exploring other areas of atomic research (sometime in the future), provided that these developments did not prevent the secondment of key scientists to serve in the United States.
6. The U.K. technical staff would work in all parts of the U.S. production establishment. The United Kingdom would therefore acquire the totality of information and experience.
7. The U.K. weapons group working in the United States would be free to transmit information to its U.K. counterpart.
8. The United States would provide highly enriched U-235 to the United Kingdom in exchange for plutonium produced in the United Kingdom.
9. About twenty atomic weapons would be stored in the United Kingdom for its immediate use. (This item represented the United Kingdom's

position. The storage of weapons issue was left open to further discussion.)[78]

The British delegates were pleased with the result of the talks. The ambassador, Sir Oliver, cabled London, in advance of the delegates return, that he, Cockcroft, and Penney felt that the proposals were "not objectionable in principle and indeed offer a mutually advantageous arrangement."[79] It was up to London to reject or accept the U.S. proposal. Within days the Ministry of Supply announced publicly that it was halting work on the third atomic pile because of "possible developments in the near future."[80] Speculation ran rife in the U.S. media that an Anglo-American-Canadian atomic cooperation deal was imminent. However, London still had some concerns. Ministers agreed that the overall arrangement offered substantial advantages to the United Kingdom. Nevertheless, two main items concerned them. They felt that the United Kingdom should have the freedom to pursue, at any time, any new processes connected with the manufacture of atomic weapons, provided that these developments did not prevent the secondment of U.K. scientists to the United States. They resented any restrictions based on duplication. In addition, the ministers were not satisfied with the American assurances pertaining to the creation of a stockpile of atomic weapons in the United Kingdom. The American delegates had been somewhat noncommittal on that item. The British feared that the question would be sent to the U.S. Chiefs of Staff, where an unfavorable decision could result. Moreover, they were cognizant of the congressional hurdle and Defense Secretary Johnson's opposition that still loomed ahead. In fact, at a December 1949 meeting, in Washington, of the Overseas Writers Club—U.K. writers had been excluded—Johnson boasted that his intention was to ensure that the British would not make atomic bombs and that, even if they agreed not to, they would be deprived of full information on American development. Several friendly correspondents at the meeting informed the British embassy of his comments and added that they would be "suckers" to fall for the sort of arrangement he had in mind.[81] Therefore, to guarantee the most binding assurances possible, nothing less than a formal document approved by the president seemed reasonable to London.[82]

In the end, the government presented a detailed counterproposal that addressed its immediate concerns and just about every other question that might arise in the future. It was a draft that left nothing to vague interpretations but drew from the memory of past experiences. With regard to U.K. scientists serving in the United States, the British draft specified a term of two to three years, after which other staff members would replace them.[83] The United States was quick to observe that under such a procedure British scientists would be in a position to accumulate American scientific methods, which they could then apply to the British program. Hence, after an abeyance of three years, the British project would sharply accelerate at the expense of the Americans. The British also reserved the right to add more production piles in the future, contingent upon the availability of raw material. The item dealing with the scope of their program was rewritten to eliminate all restrictions. And full cooperation was itemized to cover all phases of the bomb project. The remainder of the United Kingdom's stockpile of weapons, in excess of the twenty to be stored in the United Kingdom, would be stored in Canada at the United Kingdom's disposal.[84]

Ambassador Franks presented Acheson with a copy of the British counterproposal on December 30. Acheson in turn circulated copies to members of the State Department, Department of Defense, and AEC. Later, Arneson informed Franks that a considered reply would take a while, since the aforementioned departments were busy studying the proposal "with critical and in some cases suspicious eyes."[85] Despite Margaret Gowing's observation to the contrary,[86] it seemed that a compromise agreement was possible. The benefits were too great for either side to resist a compromise. Besides, they were separated by a bridgeable gap. Arneson and Adrian Fisher, legal adviser of the Department of State, admitted to Acheson that "the British position is very near the position which we were authorized to explore."[87] Both men thought that a resolution of the differences was obtainable. Moreover, the British were already exhibiting some flexibility. Their position had undergone some changes, from insistence on two mutually independent programs to acceptance of the idea of semi-independent but integrated programs. The United States in turn had shifted its position from a total ban on U.K. bomb production to a restricted program of nonduplication of U.S. efforts and full secondment of U.K. scientists to the United States.

Both nations were willing to compromise, and there were areas where compromises were possible. The number of years of secondment was a possible area of compromise, since the years of commitment to the United States could ultimately determine the British bomb production schedule. In late November 1949, the Americans had indeed convinced the British that it was practically impossible for the U.K. scientists to work on both projects simultaneously.[88] Ironically, the British acceptance of that observation might have led them to include in their counterproposal a time stipulation to guard against robbing the U.K. plants of their qualified scientists for an extended period of time. Nonetheless, Fisher and Arneson believed that the United Kingdom and United States could come to a workable compromise on that issue.[89] Satisfying the United Kingdom's request for a stockpile of weapons was not an insurmountable task either. Various American personnel had bandied the stockpile question about long before the British embraced it as part of their proposal. At the first Blair House meeting of July 14, Senator Vandenberg, supported by McMahon and Tyding, had raised the same matter that Acheson at the time had shot down. On another occasion even Hickenlooper, as mentioned earlier, expressed a willingness to make weapons "available in time of need," albeit in exchange for the nonproduction of atomic weapons by the United Kingdom. Could the administration have persuaded him and other members of the JCAE to locate those weapons in the United Kingdom under a sort of joint control arrangement? In the wake of the Soviet detonation, such a development was not unlikely. Nor was it reaching to expect the British to accept a dual-control arrangement. An American presence in the United Kingdom, to train the British in the fundamentals of the U.S. bomb, was unavoidable. In the immediate future both teams, by necessity, would have had to work together. The transfer of actual control was negotiable.

Meanwhile, the Americans worked feverishly to find an acceptable solution. In early January 1950, the administration informed the United Kingdom that it was in the process of drawing up a set of counterproposals that it planned to submit to the JCAE before presentation to the British.[90] Could the British have been persuaded to compromise even further by agreeing to a brief postponement of their bomb production effort by extending the years of secondment? Indeed, they desperately needed a stockpile, and a British-manufactured stockpile was not foreseeable. However, before the Americans could tender their own counterproposal and before any of the raised questions could be addressed, a new spy scandal rocked the boat of collaboration. As the two countries struggled to recover from the effect of that scandal, they were hit by another and yet another spy scandal. The effect that each of these spy episodes had on Anglo-American atomic cooperation would be quite revealing, as will be seen in the next chapter.

In sum, the year 1949 was a productive year in the crusade for atomic collaboration. Both the United States and the United Kingdom approached the matter with new resolve and painstaking attention to the details of any new agreement. No one wanted a repeat of the uncertainties of previous years; hence, on the part of the United States, despite the confrontational mood of the president, care was taken to work with Congress. As in the past, Congress continued to exercise its mandate to limit the proliferation of the bomb. Its members reined in the State Department and the AEC, who seemed to them a bit too generous with U.S. atomic information. Even after the Soviet explosion, although a bit subdued, the JCAE did not relinquish its prerogative. Of course, this willingness to cooperate was nurtured by the overriding consideration of a broader approach to U.S. foreign policy and by the ever-present shortage of uranium. These two factors forced the State Department to reappraise its atomic energy relations with the British. For their part, the British were all too happy to work with the Americans. In fact, they had waited patiently for months as the State Department attempted to pave the way for a new agreement. However, when the exploratory talks opened, they stood firm in their resolve not to capitulate before the demands of the Americans. They were not willing to accept cooperation at any cost, insisting that the United Kingdom had to have all the facilities needed to manufacture its own bombs. The British government was not blind to the atomic threat of the Soviet Union. Clearly, the Soviet detonation had forced them to reconsider their initial response to the Americans' proposal for an atomic agreement. An atomic stockpile in the United Kingdom was a must, even if the government had to accept an integrated program. By the end of January 1950, an agreement seemed to be in sight.

NINE

Spy Scandals Affect
Cooperation: 1950–1951

The year 1950 opened with one British spy scandal and ended with another. Before Anglo-American atomic cooperation recovered from the effect of those two scandals, a third rocked the atomic community in mid-1951. After the first two, the United States was still prepared to engage in some form of cooperation, but the third had a most damaging effect on the future of atomic cooperation. The compelling reasons that had forced the United States to reconsider closer atomic ties with the United Kingdom in 1949 did not evaporate with the discovery of Soviet spies. In fact, in 1950 the U.S. administration took steps to expand its atomic program, thereby making the reasons for collaboration even more critical. Of course, all the spy scandals adversely affected atomic relations, but unquestionably the Donald Maclean–Guy Burgess scandal of 1951 descended like the proverbial last straw on the back of the collaboration effort. The Klaus Fuchs and Bruno Pontecorvo revelations were untimely, coming at times when breakthroughs seemed imminent. In both cases the lost momentum was regained after several months, but only to succumb to a new scandal. However, the Maclean-Burgess effect lingered on for years. Anglo-American cooperation never fully recovered until some years after the British had developed their own atomic bomb.

It will be recalled that the year 1950 opened with the Americans considering the British counterproposal of December 1949. The British had agreed to an integrated program that included the transfer of U.K. scientists to the United States for a limited duration. However, they had held steadfast to their demand for a semi-independent[1] British program capable of producing atomic weapons. The demand was neither unreasonable nor hopeless. For their part, the Americans had embraced the advantages of introducing British scientists "with their new ideas and fresh point of view"[2] into the U.S. program. Furthermore, the United States remained sensitive to the potential drain on raw materials if the British were to set up a competing program in England. An effective partnership would provide a reliable source of uranium, keep the British happy, and increase the collective strength of both countries through the production of more and better weapons. Moreover, atomic collaboration would fit in nicely with Anglo-American foreign relations in Europe and across the globe. The U.S. administration was just about to hold general discussions among the Atomic Energy Commission (AEC) and the Departments of State and Defense when the Klaus Fuchs matter broke.

On January 24, 1950, Fuchs confessed to British intelligence that he had transmitted atomic information to the Soviet Union during and after the war. At the time of his arrest, Fuchs headed the Theoretical Physics Division of the British Atomic Energy Research Establishment at Harwell. It turned out that Fuchs, a naturalized British citizen who had migrated to England from Germany in 1933, had long held communist sympathies that were well known to the British Security Service. Nonetheless, since theoretical physicists were in short supply during the war, the British, doubting the veracity of their Gestapo informant, enlisted him in their atomic program. The British claimed that three subsequent security checks had unearthed no additional adverse information. John Costello, however, argued in *Mask of Treachery* that Fuchs had continued to associate openly with communist colleagues in Britain and had made no secret of his beliefs. Despite this, Fuchs was among the team of British scientists who were transferred to the United States in December 1943. He was one of those who had worked at Los Alamos, from December 1944 to June 1946, where he was intimately involved with the development of the bomb. On his return to the United Kingdom, he continued his association with the British bomb project.[3] Thanks to the relentless efforts of Robert J. Lamphere, an FBI counterintelligence agent who was privy to an intercepted Soviet decrypt containing references to the gaseous diffusion process, Fuchs was identified to the British as the possible leak.[4]

The Klaus Fuchs spy scandal could not have come at a more inopportune moment for prospects for Anglo-American atomic cooperation. It came at a time when the United States was caught up in the phobia of anticommunism. Much of the American public had already come to believe that communist infiltrators were lurking everywhere. And they were soon led to believe that subversive elements had penetrated even the close circle around the president. As early as 1947, the Republican-controlled House Un-American Activities Committee (HUAC), which was established by Democrats in 1938 to weed out unwelcome foreign influences in the United States, set out to prove that the Democrats had allowed communists to run amok in the country. The committee at first directed its attack against the movie industry, which was subjected to highly publicized inquisition-like committee hearings. Those whose loyalty came under suspicion were blacklisted from the industry, their careers and reputations destroyed, in some cases irreparably so. The most publicized target of HUAC's investigations was, in the summer of 1948, Alger Hiss, a high-ranking member of the State Department, who was accused of spying for the Soviets. Since the statute of limitations had already expired, the court charged him with perjury for denying under oath that he had held membership in the Communist Party.

The activities of HUAC sowed the seeds of suspicion in the minds of many Americans, who believed that communists had infiltrated not only the federal government but also state and local government. Something had to be done. Throughout Washington, government officials joined the struggle against the spread of communism. To counter Republican attacks, President Truman, in 1947, tried to seize the initiative by launching a widely publicized hunt of his own. He issued an executive order to review the loyalty of federal employees. Those who were deemed security risks were dismissed or transferred to less sensitive areas. By 1948, the Justice Department and the FBI had joined the hunt by targeting "dissident" organizations and radicals for investigation. By the time 1949 rolled around, most Americans shared anticommunist senti-ments. These sentiments grew with the "loss of China" to Mao Zedong's com-munists in 1949, the Soviet detonation of an atomic bomb—an achievement that many suspected had been made possible with the help of traitors—and now the Fuchs scandal, which seemed to confirm that suspicion. American reaction was quick. The Democratic-controlled Congress passed the McCar-ran Internal Security Act, which forced all communist organizations to regis-ter their groups with the government and forbade communists to travel abroad or work in defense plants. Moreover, the government reserved the right to deny entry visas to undesirables from foreign countries.

In the same month as Fuchs's arrest, Joseph McCarthy, a Republican first-term senator from Wisconsin looking for a topical issue to help his reelection campaign in 1952, claimed that 205 known communists were currently work-ing in the State Department. Although he was never able to prove the charges, for the next four years he successfully exploited the anticommunist issue to build himself a national reputation. His senate subcommittee publicly investi-gated one government agency and overseas embassy after another, looking for communist influence. So popular, powerful, and feared did he become that cabinet members, generals, and senators cowered before him and his commit-tee. Millard F. Tydings of Maryland and a few other senators who had dared to criticize him were defeated in the 1952 elections after McCarthy cam-paigned against them. Even presidential candidate Dwight D. Eisenhower thought twice about defending former secretary of state and defense and World War II hero George Marshall when he too came under attack in 1952.

The Fuchs affair had other consequences as well. Julius and Ethel Rosenberg were arrested as Soviet spies in July 1950. Fuchs had identified Harry Gold, a Philadelphia chemist, as a contact man for the Soviets. He in turn fingered David Greenglass, a U.S. Army machinist who had worked at Los Alamos. Greenglass turned in Julius and Ethel Rosenberg, his brother-in-law and sister respectively. The couple were accused of masterminding a spy operation that passed atomic secrets to the Soviets. Despite their protestations of innocence, the courts found them guilty and sentenced them both to death on April 5, 1951. For our purposes, however, Anglo-American atomic collaboration was the biggest casualty of the Fuchs affair. Secretary of State Dean Acheson summarized the results in the following terms: "The talks with the British and Canadians returned to square one, where there was a deep freezer from which they did not emerge in my time."[5] His time ended in 1953.

Surprisingly, all things considered, almost immediately after Fuchs's arrest and confession, steps were taken by the British to salvage the talks. The mutual benefits were too great to surrender without a struggle. Secretary of Defense Louis Johnson's initial response was predictable. He surmised that the information provided by Fuchs to the Soviets could give them an advantage in the race for the hydrogen bomb. Hence, he recommended to the president an immediate "all-out development of hydrogen bombs," an opinion supported strongly by the Joint Chiefs of Staff.[6] Also, he suggested to Acheson that the British be excluded from any future negotiations for South African ores.[7] Acheson rejected the idea. Instead, he proposed that the American members of the Combined Policy Committee (CPC) convene to consider all pertinent problems relating to tripartite relations.[8] The Americans met on April 25[9] and agreed to take the British up on an offer to discuss the "comparability of standards of security" existing in Britain and the United States in the field of atomic energy.[10] The security conference was the brainchild of Carroll Wilson, the AEC's general manager. He had conveyed to the British his fears that, in light of the Fuchs case, Congress would not accept any tripartite arrangement unless it was satisfied that British security standards were comparable to those of the United States. He thought that the initiative for such a meeting should come from the United Kingdom.[11]

The first of two conferences was held on June 19 through 21, 1950. (The second was held one year later in the wake of the Maclean-Burgess defection.) The conferences enabled U.S. and U.K. officials to better understand each other's security screening process. Under the U.S. system, the FBI conducted the investigation of an individual, placing emphasis on a Personnel Security

Questionnaire (PSQ). The PSQ contained questions on the individual's personal background, education, past employment, membership in listed organizations, and so on. On the basis of the PSQ, the FBI conducted a full and open investigation that involved numerous interviews. The findings, with no comments or recommendations, were passed on to the AEC, where its Division of Security made the clearance decision. The entire process took about six to eight weeks. The British process, on the other hand, took about two weeks. It consisted of a discreet background check in which the individual concerned was not an active participant and, for that matter, might be totally unaware that he was under investigation. Scotland Yard consulted its files and asked the local police to make discreet inquiries. The Security Service then summarized the report and passed on what it regarded as relevant to the atomic agency. Under both systems, judgment calls had to be made in cases where the nature of adverse information was so vague as to inspire legitimate doubt. In such cases the British granted clearance on the basis of the value of the individual to the project.[12]

The United States was not impressed with the "discreet" approach of the British system. Nor was Attlee enthusiastic about the U.S. system, viewing it as an encroachment on the liberty of the subject. He doubted its effectiveness in weeding out security risks, citing Soviet infiltration of the U.S. system.[13] For the most part, the United States found that both security standards were comparable—the United Kingdom's was no worse than that of the United States. However, it believed that British officials should question individuals more stringently about their communist affiliation. In the short term, the effect of the security meetings was minimal. The Labour Party was voted out of office in October 1951 before it could tinker with the existing process.[14]

Meanwhile, administrative changes were taking place in the AEC. David Lilienthal retired in February 1950. Sumner Pike barely won reappointment to the AEC, effective July 1, 1950. Some members of the Joint Congressional Committee on Atomic Energy (JCAE), fearing that the president might nominate him to be the new AEC chairman, opposed his reappointment.[15] Both he and Lilienthal had spoken out against the hydrogen bomb, which the JCAE and the Defense Department supported. Pike did not get the chairmanship. The president nominated Gordon Dean, who had close ties with Senator McMahon, as the chairman. Within weeks Carroll Wilson, another friend of the British, resigned in protest over Dean's nomination.[16] The British viewed these developments in an ominous light. Their interpretation was that the AEC had lost its independence and was now a tool of the JCAE.[17] But Dean's nomination did have some advantages from the administration's and, ultimately, the British perspective. He was indebted to both the president and McMahon, who had used his influence with the president to get him the chairmanship despite Lilienthal's opposition. Moreover, he enjoyed a cordial working relationship with the full JCAE, something that Lilienthal had never had. Hence, if Dean could be won over to the administration's agenda, the prospect of gaining the JCAE's support was much improved. There was another significant change. The British were pleased when the president appointed their old friend George Marshall to replace Johnson as secretary of defense in September.

Almost immediately after Fuchs's arrest, the Americans and the British too took steps to salvage the ongoing negotiations. Within days, Ernest Bevin, the British foreign secretary, suggested that he and Acheson should downplay the ill effects of the case.[18] That was not quite possible because the media was already having a field day with the issue. The publicity gave vent to wild speculations concerning the extent of atomic espionage. The matter had to be allowed to run its long course. The British also made attempts to get the Americans to declare officially that the technical cooperation clauses of the Modus Vivendi were still in effect, despite the existing disagreements and spy problems. The AEC had no problems with such a declaration. To allay British fears and leave no doubt about the status of the Modus, the AEC sent over "an unusually large number of classified reports" to the United Kingdom in early March.[19] In turn, it requested information on the British Butex process[20] "in accordance with our mutual understanding of the Technical Cooperation Program under the Modus Vivendi."[21] The AEC also agreed to expedite the calling of a long-delayed Analytical Chemistry Conference that had been postponed pending the integration effort.[22] Both sides had come to a tacit understanding to live by the ineffective terms of the Modus in order to maintain a semblance of cooperation until they found a way around their current difficulties. The Americans needed the British and vice versa.

The British were determined not to allow the Fuchs case to derail the integration talks. On an official visit to London in May 1950, Acheson experienced this determination firsthand. Bevin invited him to his apartment to discuss the status of Anglo-American atomic discussions. Acheson was surprised to walk into a five-man British team that included Sir Roger Makins and the prime minister himself. What followed was a formal and frank discussion led by the prime minister. It was clear to Attlee that an agreement would not be forthcoming within a year. Meanwhile, the British had to decide whether to suspend their program in perpetuity or proceed in a manner that might produce future difficulties with the United States. Acheson too was not overly optimistic about a quick resolution of the existing stalemate. He was against a resumption of discussions before the upcoming elections in November. In addition, the Americans could not proceed until the AEC, the State

Department, and the Department of Defense had studied the result of the Anglo-American security standards. The British did not need much persuading. They agreed to work within the terms of the proposal that they had tendered before the Fuchs case so as not to jeopardize future negotiations.

British acquiescence, however, came at a price. They expected to receive "certain materials" from the United States, the refusal of which "would cause misunderstanding and difficulty."[23] The "certain materials" turned out to be, first, an interim allocation of 505 tons of uranium to feed their Springfields refinery. The British had not received any allocation of uranium since 1946. In the interim, they had severely exhausted their stockpile to the point where the Springfields refinery faced a shutdown or reduction of its production rate.[24] Second, they requested an export license to procure a ton and a half of a plastic lubricant called Kel-F, which they needed for their low-separation diffusion plant. The State Department later questioned whether, under the provisions of the Atomic Energy Act, it was legal to grant such a license, especially since the lubricant was destined for use in a plant that the United States felt would not be needed if a partnership arrangement were worked out.[25]

As things turned out, the Americans were just as anxious as the British to resume collaboration talks. In terms of allocation, U.S. needs were even more acute in 1950 than they had been the year before. The factors that contributed to the allocation problem were due partly to major foreign policy decisions that the U.S. administration made in 1950. On January 31, the president directed the AEC to proceed with the development of a thermonuclear or hydrogen bomb. Senator McMahon, a strong advocate, lauded the decision on the floor of the Senate.[26] This decision was made primarily to provide the United States with the means to maintain its nuclear edge over the Soviets. The administration was not willing to risk having the Soviets develop the hydrogen bomb ahead of the United States. The Fuchs episode served to confirm the wisdom of the president's decision. It magnified the threat and organization of the Soviet espionage service. There was no telling how much atomic information their spy network had accumulated.

Meanwhile, until the hydrogen bomb was developed, it had become imperative for the United States to maintain its atomic supremacy over the Soviets. It was clear to the U.S. administration that its containment policy needed reexamination in the light of Soviet atomic progress and communist advances in Asia and Eastern Europe. The result was a National Security Council report (NSC-68), a draft of which was presented to the president in April. The report addressed not just the containment of communism but a program to win the Cold War and destroy the Soviet system. To that end, NSC-68 called for a major buildup of American military power. (Acheson believed that the military budget could be safely increased from $13 billion to $50 billion per year.) It urged the development of nuclear weapons as the centerpiece of American military power. Moreover, it projected large conventional forces for use in a limited war against the Soviets. The United States would use that power in a moral crusade to defend freedom around the world. The administration anticipated a fight with fiscal conservatives over the large tax increase required to finance the program. But before congressional opposition could materialize, North Korea attacked South Korea on June 24. Containment had suffered another blow. NSC-68 became the cornerstone of U.S. foreign policy for the next twenty to thirty years. The arms race was underway.

On July 7, President Truman requested Congress to provide the AEC with supplemental appropriations of $260 million for additional atomic plants and facilities. Congress had the act on his desk by September 27. Also in July, the president agreed to a recommendation by the AEC, the Joint Chiefs of Staff, and Secretary Johnson to store non-nuclear components of atomic weapons in Britain; in an emergency, only the nuclear cores would have to be rushed to Britain. They agreed that such an action would save both planes and time when both would be in short supply.[27] McMahon and his colleagues on the JCAE were still not satisfied with America's preparedness. They wanted to double U.S. atomic production capacity. On August 2, Secretary Johnson and General Omar Bradley, chairman of the Joint Chiefs of Staff, concurred with the JCAE. The military as well was not satisfied with the current production effort.[28] Moreover, if it became necessary to use atomic weapons in Korea, contingency plans called for the production of more bombs.

It was against this backdrop that the American members of the CPC met on September 7. Acheson, Johnson, Dean, and Bradley, among others, attended. The committee agreed to grant the British request for an interim allocation of 505 tons of uranium. But since the United States planned to double its own production of bombs, it decided to put the British on notice that the United States too might seek an interim allocation in the near future. In terms of tripartite relations, it came up with a new proposal that included having the British turn over their plutonium output to the United States in exchange for a stockpile of atomic weapons. Dean was not averse to having the best British scientists work at Los Alamos "in closest cooperation" with American scientists. General Bradley supported Dean, adding that he was in favor of "all-out cooperation" in the weapons field. He believed that it was unrealistic to expect the British to close down their own bomb program.[29] In effect, the Americans came very close to accepting the British proposal of late December 1949—before the Fuchs scandal broke. A pleased Acheson informed Bevin that the negotiations were back on track. He cautioned him to handle everything "as quietly as possible" and to send over representatives whose presence would not attract "comment and attention."[30]

However, this new round of negotiations was not to be. In September 1950, Bruno Pontecorvo, a naturalized British nuclear physicist, defected to the Soviet Union. Pontecorvo was an Italian who had worked in the Paris laboratories of Frédéric Joliot. In 1943 he joined the British atomic team in Canada, where he worked under Cockcroft and alongside Alan Nunn May, himself arrested in 1946. In 1949, he transferred to Harwell and was granted British citizenship. John Costello argues that the British never screened Pontecorvo before giving him his Canadian assignment. Not until 1949 did they discover that he had communist relatives living in Paris. The FBI had already discovered some communist documents in a discreet search of his apartment. Costello states that this information was sent to the MI6 officer in Washington, Kim Philby, a yet-to-be-discovered high-ranking Soviet agent. Philby tipped off Pontecorvo and advised him to defect.[31] The British and Americans managed to keep Pontecorvo's defection secret for about a year, but the atomic officials were aware of it. Once again the momentum of the collaboration negotiation was slowed. The much-anticipated meeting between the British and Americans was put on hold until the fallout from Pontecorvo's defection could be gauged. As far as anyone could tell, any information that he might have transmitted to the Soviets added little to what Fuchs had already given them. He could be of more importance to the Soviets in the future. He was one of the few scientists with a working knowledge of the reactor technology necessary to produce lithium deuteride, a component of the hydrogen bomb.[32]

The incident did not cool the Americans' interest in collaboration. After an interim period of three months, Dean attempted to jump-start the effort. In early January 1951, he announced at a news conference that he was considering asking Congress to liberalize those areas of the McMahon Act that limited the exchange of military information. The JCAE's response was cautious but encouraging. It responded that Dean's proposal would receive "a most searching scrutiny" to ensure safeguards against leaks, such as that of the Fuchs case. In terms of security, Senator Millikin revealed that "there appears to have been some other leaks" beside Fuchs. The congressional members did not reject outright an amendment to the act, but in the words of Hickenlooper, "What we must do is to weigh the mutual advantages to each nation against the hazards of a security breach."[33]

Three weeks later, the Joint Chiefs of Staff approved a recommendation by the Military Liaison Committee that U.S. atomic bombs be exchanged for British plutonium. Over the next several weeks, extensive conversations and correspondences continued between the AEC, the Defense Department, the Central Intelligence Agency, the National Security Council, and the CPC on the subject of collaboration. The main topic centered on the language of an amendment to the act and the timing of its introduction. The Americans were not talking in terms of ending their atomic relationship with the British. If anything, Pontecorvo forced the Joint Chiefs of Staff and the Defense Department to qualify their recommendations with the rider that exchanges should be limited to "the minimum of information necessary to support that exchange."[34] Sometime later, they would realize that it would not be possible to share the bomb and still limit the exchange of technical information on the scale that they had envisaged. By May of the same year, the AEC again came out for full cooperation. Dean presented to the secretaries of state and defense a detailed statement, representing the AEC's views, on the merits of cooperating with the British and Canadians. Of course, the raw material requirements of the United States featured prominently. References were made to individual and collective security, the expertise of British and Canadian scientists, and the efficient utilization of raw material.[35] Thus, on the eve of the third spy scandal, the Americans were again poised to reexamine favorably tripartite relations.

Dean had embraced cooperation with the same zeal that Lilienthal had shown, and for the same reason: more and better bombs were needed to provide for the security of the United States. This atomic arsenal could not be sustained without a constant supply of uranium. Since the world's supply of uranium appeared to be limited, the AEC was forced to consider cooperating with those friendly powers that possessed uranium or the means to produce plutonium. The United States had already begun to recognize that the two British piles and the low-separation plant were likely to be more efficient than the Hanford piles as converters of uranium into plutonium and as efficient as the new improved ones that the United States was setting up.[36] A successful arrangement would place the entire British output at U.S. disposal. We shall return soon to the Maclean-Burgess affair and its disastrous effect on the American plans.

Meanwhile, a new atomic problem had arisen between the United States and the United Kingdom. It concerned what the British believed was their right to be consulted before use of the atomic bomb by the Americans. The Quebec Agreement, one may recall, provided for mutual agreement by both countries before use of the bomb. Indeed, there was a consensus before its use on Hiroshima and Nagasaki. The British gave up this right when the Modus was signed. British officials had agreed among themselves that since the British and Americans were allies and the common enemy was the Soviet Union, both powers would most likely accept the circumstances that would necessitate the use of atomic weapons. The Korean War, however, had introduced two new enemies in a totally unexpected arena. The allies could not agree how best to deal with one of those two powers—the Chinese.

At a press conference on November 30, 1950, in response to a question, Truman declared that the United States was willing to take whatever steps were necessary to meet the military situation in Korea and that "that includes every weapon that we have." His answer to a follow-up question stated that active consideration had been given to the use of the atomic bomb. His statement received immediate worldwide attention. A press release issued by the administration later that day to allay international fears merely poured fuel on the fire. It mentioned in part that if and when the president authorized use of the bomb, "the military commander in the field would have charge of the tactical delivery of the weapon."[37] The military commander was General Douglas MacArthur, a man viewed in some circles, especially British, as a loose cannon seeking a pretext to widen the war into China to win it back for Chiang Kai-shek and the Nationalists.

At the time of Truman's "threat," the House of Commons was engaged in a debate on the Korean situation in light of Chinese intervention. Members of Parliament had voiced the fear that Truman had lost control of General MacArthur. Truman's announcement seemed to imply that he was willing to give MacArthur a blank check to determine when and if the bomb should be used. The U.S. embassy in the United Kingdom described the debate as "the most serious, anxious and responsible debate on foreign affairs conducted by the House of Commons since the Labour Party came to power in 1945."[38] Members were afraid that the conflict might escalate into a nuclear war, drawing in the Soviet Union and exposing the United Kingdom to its atomic weapons. They were concerned that the United States had not consulted the United Kingdom in a decision that could lead to its destruction. About seventy-eight Labour members of Parliament signed a petition urging the prime minister to withdraw British forces from Korea if the United States pursued its atomic solution.[39] John Hysed, chairman of the Foreign Affairs Section in the Commons, urged Attlee to proceed to Washington "before irretrievable steps are taken." Such a visit would "inspire a wave of confidence in both the Party in the House and the Party outside."[40] At an emergency session, the cabinet agreed that Attlee should present British concerns to Washington.[41] The French government, on hearing of Attlee's plan to fly to Washington, gave the visit more of a European stamp when it suggested that French Prime Minister René Pleven and Foreign Minister Robert Schuman meet Attlee in London before he left. That meeting took place on December 2. The French too expressed their consternation that the Americans would consider the use of the bomb in Korea and that the decision would be a unilateral one.[42]

Despite the big ballyhoo in London about Truman's "threat" to use the bomb and the publicity surrounding Attlee's dash to the United States, the bomb was the most fleeting issue discussed in Washington. It was not even mentioned on the agenda of the meeting. It appears that Attlee seized upon Truman's bomb statement to get a much sought-after meeting with Truman. Since August he and the Foreign Office had been seeking a summit to discuss the Far East and Western European defense. Bevin had sought a summit because he saw, as one of its benefits, an increase in the prestige of the prime minister abroad and at home,[43] where it was particularly needed. Ambassador Franks had cautioned strongly against an August summit on the grounds that the public's expectation of a momentous announcement, in the light of the Korean War, would not be met.[44] Thus, the result of a summit would be anticlimactic. The Americans too did not believe that the August timing was propitious. It does seem also that Attlee's trip was undertaken to appease public opinion and restore some semblance of unity to his restive Labour Party. Sir Oliver Franks confessed to Acheson that he thought that the principal considerations in Attlee's mind "was his domestic political situation and the increasing anxiety in the public opinion over recent developments."[45]

The December summit was an opportunity to address the agenda that had been shelved in August. The atomic bomb question was of peripheral importance to the larger Far Eastern question. The Far East dominated the six meetings that were held on December 4 through 8. Exchanges centered primarily on the British attempt to persuade the United States to negotiate with China, to support the admittance of China to the United Nations, and to discuss with China the future of Formosa and Korea. Truman had no interest in achieving a compromise with the Chinese, whom he viewed as Soviet puppets. Moreover, he could not risk another round of attack from Chiang Kai-shek's supporters in and out of Congress.[46] Attlee won no concessions from Truman, to whom he did promise British support in Korea, even before clearing up the issue of Truman's atomic bomb statement.

Attlee finally broached the subject of the atomic bomb at the end of the last scheduled meeting. Both men were alone in Truman's study while their respective aides were drawing up a communiqué. The president assured Attlee that he had no immediate plans to use the bomb; moreover, he reaffirmed that he would not use it without prior consultation with the British. They then added "prior consultation" to the communiqué. Only after Acheson alerted both men to the uproar that would erupt in Congress when the promise of "prior consultation" became public did they agree to leave it out of the communiqué.[47] Arneson would later state that he had seen a version of the communiqué where the promise of consultation was made, before Acheson and Lovett stepped in.[48] This is significant because Truman would deny later that he had ever made such a promise. The final version, however, fell far short of consultation. It stated that the president told Attlee that he would keep him "informed" of any developments that might necessitate the use of the bomb.[49]

On his return to London, Attlee told his cabinet that the president had assured him that the atomic bomb was the joint possession of the United States, United Kingdom, and Canada and that he would not use it without prior consultation. He reported to Parliament that the issue concerning use of the bomb had been settled to the government's satisfaction. Parliament and the country were relieved. But was the matter completely resolved? The president had already emphasized on November 30 that only he could authorize the use of the atom bomb. It was up to him to decide whether the allies should be consulted or informed before or after its use. The Joint Chiefs of Staff and the secretary of defense were responsible for advising the president on the military desirability of using the bomb, while the secretary of state advised him on the political effects of its use. Having made his decision, the president then would direct the AEC to turn over custody of the necessary bombs to the Department of Defense. There was no provision in this chain of decision making for any input from outside powers.

The issue of consultation was hardly settled. In the mind of the prime minister, the understanding that he would be consulted was clear, even though it depended on no written agreement. Indeed, the British delegates had made it clear to the Americans that their record held in London would include the understanding given in the original copy of the communiqué.[50] The following month the State Department made its position clear: the final communiqué revoked and superseded the original consultation promise. There was nothing the British could do short of publishing exactly what had transpired at Washington. That route, as Makins admitted, would have embarrassed the president and Acheson, thereby forcing them to deny the British version.[51] Another option was to quietly negotiate the issue. This negotiation dragged on for nine months.

Meanwhile, a disturbing development, from the Labour Party's perspective, materialized after the summit. Labour appeared to lose its will to resist U.S. pressures. It had already lost its national confidence due to its narrow margin of victory in the election of February 23, 1950. The government had survived with a slim five-seat majority. This loss of its national mandate seemed to erode Labour's international vitality. On December 14, the cabinet agreed to West German rearmament—this after Bevin, Stafford Cripps, and

Hugh Dalton, among others, had previously opposed it. Resistance usually centered on the arguments that it was a needless provocation to the Soviets, that the French would feel threatened from a rearmed Germany, and that the German Social Democratic Party itself was against rearmament. The cabinet allowed the United States to convince it that the security of Europe hinged upon an armed German buffer sitting between Western Europe and the Soviet Union. Again, on January 25, after a bitter bout of cabinet infighting, the cabinet voted an increased defense budget of 4.7 billion pounds, from 3.6 billion pounds. The U.S. Chiefs of Staff had been pressuring them for an even higher increase of 6 billion pounds. Attlee then warned Britons that austerity would be tightened to guarantee enough materials and labor for the defense program.[52] In January 1951, the United States continued to exert pressure on the United Kingdom by insisting that it support a UN resolution branding China an aggressor nation. Fearing an escalation of the war, the cabinet was unwilling to support the resolution. It was equally unwilling to oppose the United States. Fortunately, Israel took the government off the hook by sponsoring a last-minute amendment. The U.S. government continued to make increasing demands on the British. The United States, it seemed, viewed the British as silent and obedient allies who were expected to accept its senior leadership.

The British exercised no control over American military planning, even where the use of the atomic bomb and American warplanes from British airfields were concerned. American bombers, and almost 17,000 American personnel, had been stationed in the United Kingdom for nearly three years. Yet not until January 1951 did Bevin see fit to ask the Americans to share their strategic air plans with the British. And this request was made only because Labour feared that Winston Churchill, leader of the opposition, would raise the question in Parliament. The government wanted to be in a position to reply that they were acquainted with American plans.[53] Air Marshal Sir John Slessor, British Chief of Air Staff, and Air Marshal Arthur Lord Tedder, chief of the British Joint Services Mission in Washington, were subsequently given "such information about plans as [General Omar Bradley] considered appropriate."[54] The British were given just enough information to meet the parliamentary situation.

The Labour Party's atomic policy came under persistent and intense attack from the opposition Conservative Party. On December 14, 1950, Churchill voiced his astonishment over the United Kingdom's failure to develop an atomic bomb.[55] On January 30, 1951, he forced the prime minister to agree to ask the United States whether there were any reasons why the Quebec Agreement should not be published.[56] Clearly, Churchill's motive was not selfless. His party could benefit from the publication of the agreement. It would show the British people the quid pro quo arrangement he had worked out with Roosevelt, only to have it sabotaged by the Labour Party. Attlee did make the request, but the State Department and Department of Defense strongly opposed publication on the grounds that (1) the agreement did not represent the present tripartite arrangements, (2) it would raise questions about the present arrangements that they were not prepared to answer, and (3) publication would generate questions and discussions that might jeopardize the chances of a new agreement.[57] Churchill was not deterred. He sent a personal request to the president asking for publication.[58] Needless to say, this lack of protocol on the part of the leader of the opposition raised some eyebrows in London and Washington. Attlee was livid. He did not understand the "basis" of Churchill's communication.[59] After an internal debate concerning whether and how Truman should respond, the president cited the aforementioned reasons for rejecting publication.[60] In July Lord Cherwell got into the act when he rose in the House of Lords to criticize the slow development of the British atomic program. He suggested that the program be transferred from the slow bureaucracy of the Ministry of Supply to a more flexible special organization. "Why is it," he asked, "that the Russians with their imperfect information have been able to produce a bomb in half the time it is taking us?"[61] Over the forthcoming months Parliament would continue its attack on Labour's atomic and foreign policy.

Labour's foreign policy, in particular, seemed to be floundering. The weakness of this policy eventually damaged atomic collaboration. The Americans questioned the value of integrating their program with that of a declining power. Foreign policy setbacks, including security breaches, caused the British to lose any chances of an equal nuclear partnership with the United States. On March 9, 1951, Sir Herbert Morrison replaced an ailing Bevin as foreign secretary. (Bevin died a month later.) The Conservatives had a field day criticizing Morrison's inexperience and lack of coherent policy. The prime minister himself confided that the appointment was a mistake. "His ignorance was shocking. He had no background and knew no history."[62] Acheson too complained about Morrison's ignorance of foreign affairs and his abrasive and quarrelsome nature.[63] Acheson knew of the confusion in the Foreign Office. Sir Oliver, with whom he had by then developed a very personal relationship, kept him informed of the problem "of inertia and exhaustion in London and lack of leadership."[64]

This information was fully exploited by the United States. On May 2, 1951, the Iranian government announced the nationalization of the British-owned Anglo-Iranian Oil Company at Abadan. Morrison pressed the cabinet for a military intervention. Emanuel Shinwell, the minister of defense, warned that giving in could lead to the nationalization of the Suez Canal and the collapse of British power throughout the Middle East. Lord Fraser, the First Sea Lord, recommended an invasion. However, the United States, rivals with the British for control of Middle Eastern oil, brought intense pressure to bear upon the British to refrain from military intervention. Acheson urged a peaceful solution. The cabinet rejected the military option, and by October 4 all British personnel were evacuated.[65] The Americans replaced them. The withdrawal did not improve British prestige in the Middle East. Considered in the context of earlier retreats from Greece, India, and Palestine, Abadan was perceived as yet another example of waning British influence and a decline in the British will to resist when challenged. This weakness was not lost on the United States, which had no doubt that the Labour Party would dance to any tune it played. Other than for its own national interest, the United States had no reason to make any foreign policy concessions to the British. On October 8, there was another humiliation. Nahas Pasha, the prime minister of Egypt, announced the unilateral abrogation of the 1936 Anglo-Egyptian treaty, through which the Egyptians had accepted a British military presence in the Suez Canal Zone. Morrison again rattled his saber, and again the United States protested. Attlee came out against military force, but before he could come up with a solution, he was voted out of office on October 26.

Four months before the Labour Party was voted out, still another spy scandal broke in June 1951. The British announced that Donald Maclean, head of the American Department of the Foreign Office, and Guy Burgess, second secretary of the British embassy in the United States from August 1950 to May 1951, had disappeared on May 26. They resurfaced sometime later in the Soviet Union. The government downplayed both men's importance as sources of classified information. However, Maclean had served in Washington from 1944 to 1948, where he had access to secret diplomatic documents flowing between Washington and London. Moreover, from January 1947 to August 1948, he had served as the British secretary on the CPC, where all areas of Anglo-American atomic relations were discussed. As head of the Foreign Office's American Department in London, he was privy to valuable information on the Korean War, which, undoubtedly, he transmitted to the Soviet Union. Again, the FBI played a role in uncovering him. By the end of 1949, the bureau was convinced that the Soviets had received information from the British embassy. The FBI informed the British, who took about a year to zero in on Maclean. Burgess had served in Washington for only a few months, but he too had been providing diplomatic information to the Soviets before and after he arrived in Washington. Like Pontecorvo before them, Philby tipped off both men. The fallout from this third spy scandal shattered any hope for atomic collaboration.

The Department of Defense and the Joint Chiefs of Staff had had enough of British security breakdowns. They retracted their position of January 31, 1951, the so-called Kiefer Plan, which had called for the exchange of plutonium for bombs. It will be recalled that the plan had included the stipulation that exchanges be limited to a "minimum of information." Defense had hoped to limit the British only to information that would assist them in producing plutonium. Subsequent discussions with the AEC had disclosed that it was impractical to restrict the disclosure of technical information to the limited area of plutonium production. The Defense Department was not willing to risk an expanded transfer of "significant information on U.S. atomic weapons" to the United Kingdom. In the words of the memorandum, "'Leakage' of sensitive atomic information is the most damaging single factor in the present determination of US military posture. Our present risks are matters of daily concern, and any action that tends to increase these existing risks affects the security of the US."[66] The exchange of atomic information, the statement suggested, should be placed on hold. Beyond a doubt, the military men's decision was partly influenced by the persistent breakdown of British security. Simply put, they could not trust the British with sensitive atomic information.

Two other factors, however, also served to influence the decision. The weakness of the British government played a role. By 1951, the American government had little respect for its counterpart in London. The Labour Party had become a subservient puppet of the United States. If the international community had any doubts about the decline of Britain as a leader of the free world, these doubts were quickly removed after 1950. The United States harbored no delusion that the United Kingdom still possessed the status of a great power. She was a declining power not worthy to share America's atomic information. To gain equal atomic status, the United Kingdom had to prove its worthiness by, as the Department of Defense put it, developing its atomic facilities to a point where it would be able to "fabricate a considerable number of atomic weapons."[67]

Perhaps the most important factor influencing the American changing attitude toward the British was the uranium situation. As demonstrated on numerous occasions, there had always been a clear connection between uranium shortages in the United States and the renewal of efforts to achieve Anglo-American cooperation. Not even the first two spy scandals could derail tripartite negotiations because the U.S. atomic program needed uranium. However, the third scandal occurred at a time when the U.S. uranium supply situation was about to improve. The Americans knew that future supplies beyond Britain were on the horizon. The Redox system for recycling irradiated uranium ore was about to pay dividends. At long last, in August 1951, General Electric began operations at its new Redox plant at Hanford. In addition, other companies, including Kellex Corporation, were experimenting with other processes.[68] The future of recycling was encouraging.

The increased output from U.S. mines and the potential output of international sources also gave U.S. officials reasons to be optimistic. By December 1950, the Colorado Plateau deliveries had exceeded the 1950 forecast by 60 percent, and a greater quantity was expected from new deposits near Grants, New Mexico. Some phosphate beds in the West and in Florida also seemed attractive. Overseas, a uranium source was found at Radium Hill in South Australia. The premier of Australia sounded out American interest in the deposit in August 1951, and steps were underway to establish a special relationship with Australia.[69] Negotiations were also conducted with General Francisco Franco, the Spanish leader, who was encouraged, despite his pessimism, to accelerate the search for uranium in Spain. However, the AEC still viewed South Africa as the United States' best potential source. Its first plant was due to go into production by March 1952, and full production was expected by 1954 to 1955. The AEC speculated that South Africa could produce "as much ore as the Belgian Congo, probably more than Canada and undoubtedly more than the US."[70]

How did Canada fit into all of this? For several years the Canadians had been mining a limited amount of ore. By early 1951 they had made a substantial discovery in the Athabaska region, a discovery for which Gordon Dean believed it was worth amending the McMahon Act. The Canadians had no desire to ship the raw ore to the United States. If the Americans wanted the uranium, they had to allow the Canadians to construct a refining plant using the American process, which was far more efficient than the current Canadian process.[71] To do this the AEC would have to disclose restricted refining data to the Canadians. This disclosure could not be made without an amendment of the McMahon Act. The Americans thought highly of the new Canadian source, surmising that, in time, it could surpass Congo production. The president gave his approval for Dean to take up the matter with the JCAE.[72]

The fact remains that the Americans believed that the problem of uranium shortages was about to end and that therefore there was no need to continue stroking the British. Rather, the British had become dispensable in the uranium equation. The U.S. Defense Department's decision to take the Kiefer Plan off the table was not well received by the British, but they suspected that factors other than the spy scandals were involved. One memorandum speculated as follows: "Perhaps (though this is guesswork) they [Defense] felt that the US programme was now technically so far ahead of ours, especially in the development of new types of atomic weapons, and that their stockpile of fissionable materials was large and increasing so rapidly, that it was not worth their while to try for closer cooperation with us."[73] The guess was accurate on both points, but the latter was more relevant to the current development. The Americans did believe that their stockpile and Congo allocation were sufficient to bridge the gap until the delivery of new sources of uranium.

U.S. indifference to the British continued to be reflected in other matters. Prior consultation was still an unresolved issue in September. In fact, the debate had intensified with the submission to the Americans of a British paper that examined the possible forms of communist aggression that might necessitate the use of atomic weapons. Basically, the paper formalized the argument that the British and Americans should have prior discussions concerning the respective conditions that would require an atomic response. The Joint Chiefs of Staff suspected that the British were seeking an implied commitment from the United States for prior consultation before use of the atomic bomb. Such a commitment could not be given. Acheson sympathized with British concerns. The British were in the line of fire and had no control over their destiny. As he put it to the chiefs, "They are now the tail of the kite, and they are concerned about where the kite is going."[74] But his understanding was not translated into concessions. On a subsequent visit to Washington, Foreign Secretary Morrison tried to no avail to get a prior-consultation commitment from Acheson. The prime minister expected yet another attack on consultation from the opposition in the House of Commons. What Morrison took back to London was an American promise not to use British airfields for atomic bombing without prior consultation and agreement from the U.K. authorities,[75] as well as a vague statement for public consumption drafted on the basis of the December 1950 communiqué.[76] Morrison was disappointed, but this concession was greater than any he had been able to wring from the administration in prior meetings. Moreover, it had the full support of the various American departments. The consensus was that the British deserved a voice in the decision to dispatch bombs from Britain, since there could well be bombs coming back the other way.[77]

The British were also disappointed when they approached the Americans in April 1951 for permission to use their test site in the Nevada desert to test a U.K. prototype bomb sometime in late 1952.[78] It was important for the British to know where they stood with the Americans. If the Americans refused, British intentions were to negotiate with the Australians for permission to use Monte Bello Islands off the coast of western Australia. To hold the trial in Australia would involve a major diversion of resources in terms of men, money, and material from the British defense program,[79] but that seemed the only viable alternative. The U.S. Defense Department, initially receptive to the idea of hosting the test, quickly withdrew its support after the AEC pointed out that although the United States would gain information on the U.K. weapon, it would be impossible to conduct a test without disclosing U.S. restricted data. Dean could not imagine how fifty British scientists and technicians could get together with their U.S. counterparts without the latter revealing classified information. The McMahon Act outlawed such a revelation. The law had to be amended to facilitate the test, or the conditions of the test had to be modified.[80]

Information already had been leaked to the British that the Americans were considering limiting the number of British technicians from fifty to five or six in order to control the exchange of restricted information. They were relieved when the Americans, thanks to Dean's efforts, agreed to accept the original fifty proposed by the British.[81] Nevertheless, the test would have to be conducted under very restricted conditions. For example, the proposal did not provide for full cooperation and reciprocity, since such freedom would have given the British access to U.S. restricted data. Only the Americans had the right to define what constituted restricted data. Hence, data that the British might consider essential to a satisfactory test could be denied on the basis of U.S. interpretation. Past experience had taught the British that they could not rely on U.S. lawyers to come up with objective interpretations. It was equally unclear how much access the British would be given to the U.S. laboratory and to the measurements and techniques used. For example, the Americans could decide, on the basis of their interpretation, to do most of the testing themselves and turn over the base results to the British.

The Labour government had "serious doubts about the wisdom of acceptance." Acceptance would have entailed an abandonment of the Monte Bello site, and once abandoned, it would have been difficult to get it back on track in time for the test. Rather than risk making a hasty decision, the government decided to wait until after the United Kingdom's October elections.[82] The British did not hold the American government responsible for the restrictions imposed. They knew that the United States had to work within the constraints of the McMahon Act. Until or unless the act was amended to address the exchange of restricted data, bomb information could not be shared with foreign powers.

The McMahon Act was in fact amended at the end of October 1951, but this was not done to facilitate improved Anglo-American cooperation. The amendment was hastily passed to promote American atomic needs. Ever since the limitations of the Modus Vivendi had become evident to the British, they

had been pushing the Americans for an amendment to improve atomic cooperation. Early in 1951 the Defense Department had come up with the Kiefer
Plan to exchange U.S. bombs for British plutonium. Prospects for the Kiefer
Plan looked good during the first half of the year, and it seemed as though the
U.S. administration would seek an amendment to facilitate the terms of the
plan. However, when the Maclean-Burgess case surfaced, it raised serious
doubts about the effectiveness of British security. The Americans turned their
attention away from technical atomic cooperation but not from an interest in
amending the act. The act still had to be amended to address the Canadians'
demand for American technology to construct a refinery for processing uranium. Moreover, the Canadian heavy water reactors were quite advanced. The
Americans had taken an interest in that system, but to study it required an
exchange of restricted data. The Australians too were seeking an American-
type refinery, since the cost of shipping raw ore from Australia to the United
States was prohibitive. An amendment was required to facilitate the sharing
of refinery information. In addition, the British had gathered some atomic
intelligence data on the Soviet Union that the Americans wanted. The British
were willing to give the Americans the data if they both collaborated in evalu-
ating the information and sharing the results. The McMahon Act prevented
this. Further, in the refining of uranium, British process loss was about 15
percent compared to less than 1 percent for the Americans. The U.S. officials
had to address that loss of valuable uranium. But again, the act prohibited
them from sharing their more efficient technology with the British.[83]

On September 12, Dean outlined the case for an amendment to the JCAE,
asking its members to sponsor the new legislation. The amendment was introduced on October 8 and passed on October 30. The result did little to encourage technical cooperation with the British. If anything, the amendment was
drafted so as to prevent the exchange of restricted information with countries
whose security standards were deemed not equivalent to those of the United
States. Since the United Kingdom had not yet introduced a system of positive
vetting, it was adversely affected.[84] Moreover, any exchange of atomic information would require the unanimous agreement of the AEC and the approval of
the president and the National Security Council. The JCAE had to be informed too, and no action could take place until thirty days after this. There
was also a stipulation that the exchanges must "substantially promote the
common defense and security of the United States."[85]

The British were not too optimistic that the amendment would improve
relations. Cockcroft, after discussions with the AEC, surmised that collaboration might be obtained in particular areas, such as test trials, intelligence,
and raw materials, where the United States stood to gain substantially. But

he anticipated no "improved technical collaboration in the major fields of weapon design, production of fissionable material and nuclear power."[86] He suggested that the British forget the United States and pursue their own interest in the latter fields. They should secure their fair share of uranium and make the maximum use of Commonwealth scientists and technicians.[87] They had already started along that independent path.

On the eve of the Labour government's fall, Attlee felt reasonably assured that the United Kingdom's atomic program was firmly established. The two production piles were in full operation, and the chemical separation plant was expected to begin producing plutonium by early 1952. A low-separation plant was under construction, with a projected completion date in 1953. Since the possibility of exchanging British plutonium for American U-235 had grown remote, the government had decided to build its own high-separation diffusion plant to balance fully its program for military and industrial needs.[88] Moreover, a prototype atomic bomb was expected to be ready for testing by the end of 1952. Winston Churchill was about to enter the scene to add the final touches to a program he had initiated ten years earlier.

By the end of 1951 it was clear to all concerned, on both sides of the Atlantic, that Anglo-American atomic cooperation was nonexistent in the present and unlikely in the immediate future. Even the thin fabric of nontechnical cooperation that the Modus had provided was shredded. Both governments dropped the pretense of cooperation that had surrounded it. The events of 1950 and 1951 had killed any hopes of Anglo-American atomic cooperation. The spy scandals, the Labour government's weakness, and the improved raw material situation had compelled the Americans to reappraise their ambitious integration plans. It is clear that the untimely defection of Maclean and Burgess did more harm to the collaboration effort than Fuchs's arrest. After the first two spy scandals, the Americans had still been willing to exchange bombs for plutonium, and, undoubtedly, the British would have accepted. They would have accepted because the Americans were prepared also to concede to them the continuation of an independent British program. Ultimately, this independent program would have benefited from the experience gained by those scientists seconded to the U.S. program. But all of this was not to be. Integration was abandoned because the Americans stood to risk too much in the security field and gain too little in terms of "common defense and security." In the long run, American reluctance to accept the British as equal partners forced them to stand on their own two feet and to use British ingenuity to forge an independent atomic program.

TEN

The British Develop a Bomb
but No Cooperation: 1952

On the eve of its defeat at the polls, the Labour government resigned itself to the futility of improved Anglo-American atomic energy cooperation. The prime minister and his atomic energy officials turned their full
attention to an independent project and gave active consideration to collaborating with Commonwealth and European nations. All signs also pointed to
a rejection of the U.S. counterproposal on the test site. About six months after
he took office on October 26, 1951, Winston Churchill, the new prime minister, also came to accept, albeit reluctantly, that the wartime relationship, upon
which he had been basing his optimism for improved atomic collaboration, no
longer counted for anything with anyone in Washington. Churchill was unwilling to accept that his Quebec Agreement had died with Roosevelt and was
buried with his own 1945 electoral defeat. He was sincere in his belief that
once the British proved that they had something to offer in an atomic relationship, the Americans would come around. It was in that spirit that he looked
forward to and welcomed the detonation of the British atomic bomb at Monte
Bello. He expected the Americans to embrace the British, ushering in a new
period of atomic partnership. But the detonation failed to bring them together.
Within weeks of the British successful test, the Americans exploded a thermonuclear "device," placing them even further ahead of the British. The Americans still had much more to offer than the British, and yet another secret
to guard.

When Churchill took over the reins of government, he was surprised at the
strides the British atomic program had taken under Labour. In the months
preceding the election, he had attacked the government unremittingly on the
alleged slow pace of British atomic progress and on the issue of consultation,
or lack thereof. But the British program was not as stagnant as he had believed. The outgoing government had done a creditable job in overcoming the

formidable financial obstacles and the recalcitrance of the American government to establish the United Kingdom's atomic program on a sound foundation. They were on the threshold of producing an atomic weapon, an achievement that owed nothing to American collaboration. According to Margaret Gowing, Churchill was filled with a mixture of admiration, envy, and shock when he discovered that the Labour government had spent 100 million pounds on the atomic program without the knowledge of Parliament. He felt that he would have been branded as a warmonger for a similar feat.[1] Nevertheless, he did accept the recommendation of Lord Cherwell, his paymaster-general and atomic adviser, and Duncan Sandys, the new minister of supply, to continue the policy of secrecy for fear of revealing too much to the Russians or evoking the wrath of the British people. Cherwell had remained intimately involved with the program from its inception. During Labour's administration, he had remained in touch with the project as a member of Lord Portal's Technical Committee in the Ministry of Supply.

Despite acknowledging Labour's achievement, Churchill believed that he could have done more for less. He was convinced that the Quebec Agreement would have carried more weight after the war if he had remained in office and that the United Kingdom would not have spent as much as it did to duplicate information the Americans already had. Cherwell did not agree. He thought that Churchill was unfair to Attlee for chiding him on not publishing the Quebec Agreement before the McMahon Act became law and for his failure to retain "the right to consent" clause of the agreement. He told Churchill, "If you had been in power and had heard the whole story I am sure you would have come to the same conclusions."[2] It took some doing before Cherwell could get Churchill to reconcile himself to the fact that the Quebec Agreement had been superseded by other agreements. Nevertheless, he still harbored the thought that he could resurrect it through a personal presentation to Truman. Apparently, this obsession with the past was not restricted to atomic energy. In a dispatch to the State Department, the U.S. ambassador to the United Kingdom observed that Churchill "is increasingly living in the past and talking in terms of conditions no longer existing."[3] Two weeks after gaining office, Churchill succeeded in getting Truman to agree to a January conference in Washington to discuss a wide range of issues affecting Anglo-American relations. The agenda listed atomic energy as one of the topics.

The Americans had reasons to approach the summit with some trepidation. On November 9, at the Lord Mayor's banquet at Guildhall, Churchill included the following in his speech: "It must not be forgotten that under the late Government we took peculiar risks in providing the principal atomic base for the United States in East Anglia, and that, in consequence, we placed ourselves in the very forefront of Soviet antagonism. We have therefore every need and every right to seek and receive the fullest consideration from Americans for our point of view, and I feel sure this will not be denied us."[4] The speech caught Washington's attention. Members of the Joint Chiefs of Staff and the State Department debated its ramifications. Did the "fullest consideration" mean that Churchill was after atomic information or consultation before use of the bomb? And if he did not get his way, was he prepared to evict them from East Anglia? Twelve days after the Guildhall speech, Churchill addressed part of the Americans' concerns. In response to a question in the House of Commons, he stated that U.K. bases and facilities would continue to be made available to the U.S. Air Force so long as the arrangement "is needed in the general interest of world peace and security."[5] Despite this clarification, Churchill's position created a degree of uncertainty and concern in Washington. Senator Brien McMahon, chairman of the Joint Congressional Committee on Atomic Energy (JCAE), believed that Churchill was out to blackmail the United States and was deliberately alluding to the bases to send a message to Truman. McMahon shared his suspicions with the president. Churchill did not intend to come to Washington as a petitioner, McMahon warned: "On the contrary, owing to the vital importance of the bases to us, he is in strong position."[6] The following day the bases were the subject of yet another parliamentary question concerning whether the United Kingdom had the right to terminate the base arrangement if it did not wish to be involved in a U.S. war. All of this talk about the bases created a sense of unease in the United States, leading to the perception that Churchill had planted the questions. However, there is no evidence to support any such charge. It is more likely that the parliamentarians, so long deprived of atomic information by Attlee, were hoping to inform themselves at the expense of their new prime minister, still flush with the elation of victory. The December 6 query, for example, came from Sydney Silverman, a member of the opposition Labour Party.

What happened next did not allay American fears. On December 26, acting on instructions from London, the British ambassador, Sir Oliver Franks, informed Secretary of State Acheson that the British had decided to proceed with their atomic test in Australia. The British statement was emphatic, closing the door on the chances of any further negotiations. It explained pointedly that the American offer was too restricted in that it did not provide for full cooperation and reciprocity. Moreover, the statement added, the British really preferred a shallow-water test, which could be facilitated at Monte Bello but not at Nevada.[7] This note of finality and decisiveness was not one that the Americans had come to associate with the British on atomic matters. It seemed to herald an independent approach by the British that was not lost on the Americans. Could Churchill be signaling a willingness to forge an independent atomic policy, and if so, how would the United States be affected? The president was alerted to the possible wind of change and the likelihood that Churchill might have more to say on that subject at the summit.[8]

The Americans also feared that Churchill intended to establish an informal one-on-one type relationship with Truman. Such a relationship with Roosevelt during the war had created enough problems, as they were well aware. The State Department had no intention of allowing Truman to wander into that trap and fall under the spell of Churchill's eloquence. The department was equally concerned that Churchill might not wish to be constrained by any agenda but might choose to range freely over a wide variety of subjects.[9] It advised Truman to avoid any such attempt and to insist on the presence of his advisers at all sessions with Churchill.[10] In the end, the Americans did extract an agenda from Churchill. It included topics on the European army, the Atlantic Command, the Soviet Union, the Far East, the Middle East, Southeast Asia, British economic problems, relations with Germany, the strategic air plans, the use of atomic weapons, and technical atomic cooperation. It was in fact a grand tour d'horizon, and the Americans expected Churchill to make random observations not only on these but on a multitude of other issues. They were prepared to give him the freedom to touch all the bases and outline the general policies of his new government.

In terms of atomic policy, Cherwell's advice to Churchill was comprehensive and insightful. He advised him not to press the Americans for a new amendment to the McMahon Act. Such an effort would be futile. Rather, he should focus on getting the Americans to recognize the waste of effort and resources involved in noncooperation. They should be persuaded to engage in the maximum cooperation possible under the existing law. Emphasis should be placed on the comprehensiveness of the British program and its potential for contributing to the joint effort. Above all, the British should not present themselves to the Americans as supplicants. As far as the production of fissionable materials was concerned, the British program was independently successful and required only marginal assistance from the United States. Technical cooperation, though still desired, had lost its urgency. Of greater concern to the British were the strategy and tactics involved in the use of atomic weapons and the degree to which the British would be involved in the decision-making process.[11]

Cherwell's advice was well taken. The summit met from January 5 to 11 and again from January 16 to 18 (in between, Churchill visited Canada). Atomic energy matters were introduced at the second formal meeting on January 7. Churchill repeated his argument for the disclosure of the history of the wartime relationship, as a matter of record. The statement failed to elicit a comment from the Americans, and Churchill did not force the issue. In the field of technical cooperation, he hoped to get fuller cooperation within the limits of the law. The president was quite agreeable to Churchill's suggestion that Cherwell meet with the appropriate American representatives to discuss technical cooperation. The administration was relieved that Churchill's requests were not more extreme. They were also pleased when Churchill indicated that his government was taking steps to tighten its security, not just in the atomic field, but in all areas of government where classified information was involved.[12] Acheson had made it clear that British security was still a major barrier to cooperation. With a sense of relief the two delegations moved on to other items. The British, on their part, were satisfied that the president had instructed his people to pursue the matter in further discussions.

The follow-up meeting of January 10 dashed any optimism the British delegates might have held. The meeting began under ominous signs that the Americans did not place the same degree of importance on the discussions as the British did. In attendance at the start were Cherwell and Makins, Arneson of the State Department, and two AEC commissioners. One commissioner, Thomas Murray, left early. Gordon Dean arrived after the session started. Robert Lebaron, deputy secretary of defense for atomic energy matters, and his military people walked in thirty minutes late. Dean promptly pointed out that in the present climate of Congress it would be difficult to exchange information on the broad areas of atomic energy. The exchange of information might be possible only if requests were made on specific items within the broad areas. Even so, the United States had to be in a position to determine in advance whether a specific exchange would be of substantial advantage to the United States, as required under the amendment. Thus, Dean suggested that the Americans send a team to the United Kingdom to study the U.K. program. Cherwell responded that "opinion in the UK would not want to see us opening our programme to American inspection if we were debarred from seeing the American programme." He thought that any such proposal would have to be on a reciprocal basis. Dean agreed.

It became quite obvious to the U.K. delegates that the AEC was receptive to a more liberal interpretation of the nine areas of the Modus and was willing to fit some form of collaboration under the amendment. However, the Defense Department was not as flexible. It was equally apparent to the British that they were caught in an interdepartmental power struggle. It turned out that Lebaron's inflexibility was due to the AEC and JCAE's drawing up the amendment to the McMahon Act without the participation of the Pentagon. The Joint Chiefs of Staff thus found it difficult to go along with specific proposals that were based on an interpretation of an amendment for which their advice was not sought. As far as Lebaron was concerned, the McMahon Act still forbade the exchange of military information, and, in his view, all atomic information was military information. Dean commented that personally he could not see what was troubling Lebaron.[13]

Arneson would later state that "it seems clear that the spirit and intent of the Truman-Churchill exchange has not been instilled in Mr. Lebaron." Arneson expected Cherwell to report the negative tone of the meeting to Churchill.[14] He did not. Secretary of Defense Robert Lovett strongly discouraged him from pursuing the matter at a higher level.[15] Arneson himself asked Acheson to take up Lebaron's attitude with the president and Lovett to ensure that the president's policy was carried out. In the end Acheson did not do so. Apologizing later to Arneson, he said, "I am sorry I failed to take this up with Lovett. But it is true that the President did not make an agreement. He merely expressed his view that the possibilities of fuller cooperation should be explored."[16] Thus, both leaders were totally unaware that their decision to resume discussions on technical cooperation had run into a roadblock thrown up by the Pentagon. It is quite likely that the president, had he known, would have looked into the matter. Churchill would not have been surprised to learn of Lebaron's resistance. Earlier, McMahon had identified Lebaron to Churchill as one of those who still opposed exchanges.[17]

A second meeting was held to discuss scientific atomic energy intelligence cooperation. That session was more successful. Bedell Smith, director of the Central Intelligence Agency, did not anticipate any serious obstacles in this area. His optimism was well founded, since the Americans stood to gain British intelligence data in this type of cooperation. Overall, the result of the atomic discussion at the summit was mixed. Like Attlee before him, Churchill could not get the Americans to agree to prior consultation in all theaters of conflict. The British would be consulted before British bases were used, and NATO allies would be consulted in the event of an European conflict. But in other areas of the world, the Americans could not make a commitment. Unlike Attlee, Churchill did get a personal briefing from the Department of Defense on the strategic air plan. By the end of 1952, even the British Chiefs of Staff had received, on a highly personal basis, full information on the strategic and tactical aspects of the air plan.[18] Before Churchill left Washington, the AEC agreed that Sir John Cockcroft should visit Washington to discuss specific instances of cooperation and exchange of information.

The prime minister left Washington fully satisfied over the summit. He had expressed his views on an array of topics, and the Americans had given him a red carpet and an attentive ear. American and British foreign policy, for the most part, did not diverge greatly. But if Churchill returned to London with any illusions that he had resurrected atomic collaboration, he was alone. Those who had been intimately involved with the controversy knew that the summit had changed nothing. Two weeks after the summit, the British embassy reminded London that the amendment that the AEC intended to use as the basis for future collaboration had specifically stated that the existing technical cooperation arrangement with Britain and Canada remained unaffected.[19] "Existing technical cooperation" had been nonexistent before the passing of the amendment, and it remained so after its introduction. The embassy was pessimistic over the chances that the British and the Americans would agree to future topics of exchange. "Even if we in the UK had a bright idea," it added, "it would have to be a very bright one before the USA would be interested."[20] Any exchange of information had to be of "considerable advantage" to the United States, and only the AEC could interpret that phrase. Once interpreted, approval had to be sought from the National Security Council, where the Department of Defense was represented well.

Lebaron had already taken steps to ensure that the Department of Defense exercised a veto over all exchanges even before they were presented to the National Security Council. He sent a letter to the AEC outlining the willingness of his department to cooperate with the AEC in effecting the amendment, but, he added, the Military Liaison Committee must be informed of any proposal "before the Commission took any formal action under the amendment." The AEC thought that it would be "unwise" to turn down the request.[21] Reports had also reached the British that there had been little enthusiasm from scientists in the Reactor Development Division of the AEC for closer collaboration. The Reactor Division had decided that it could not "honestly claim" that an exchange of information would be of substantial benefit to the United States.[22]

Future collaboration did not look too bright, and Cherwell and his subordinates were not prepared to delude themselves. Cherwell had no qualms about playing the Americans' game. If they insisted on being secretive, well, so would the British. London sent instructions to the Washington embassy that it should avoid discussing Hurricane, the British code word for the Australian test, with the Americans. Discussions with them "might encourage Americans to ask for information about Hurricane or to request permission to send observers . . . , which we wish at all costs to avoid."[23] A few weeks later, Churchill himself announced publicly that the Americans would not be invited to witness the test. Ironically, this stand was taken just when the Americans began volunteering information on one of their own tests at

Eniwetok. Perhaps this overture was made with one eye on reciprocal exchanges from the upcoming British test. But London was having none of it.

Cockcroft's visit in late March contributed nothing to convince the British camp that things would improve. The trip made clear to the British that the Americans had laid upon them the onus of proving to the technical officers of the AEC that any exchange of information would be of substantial advantage to the United States. To do that, they were expected to provide enough information on the British program to make a case. In the meantime, the Americans had no such responsibility to open up their program. In fact, the detailed discussions that Cockcroft held with the AEC staff were extremely one-sided. He was required "to expose the whole field of UK research and development work," whereas no reciprocal information was, or could be, provided by the AEC.[24] In the end, the only valuable result of the discussions was the AEC's decision to extend cooperation in atomic energy intelligence. Both the CIA and the Department of Defense favored that action. However, certain materials, such as helium and Kel F, which the British needed for their diffusion plant and power reactors, were not readily accessible.[25] Efforts to obtain improved collaboration in nuclear physics and on the military application of atomic weapons met with no success.

In the weeks following Cockcroft's return to London, it became painfully obvious that Anglo-American atomic cooperation anywhere in the immediate future was quite unlikely. A U.S. presidential and congressional election was approaching in November. The AEC accepted the virtual impossibility of persuading Congress to approve any additional areas of cooperation under the amendment before its recess.[26] Moreover, the AEC had no intention of making any submission to Congress until after the AEC had verified that British security standards were indeed comparable to those of the United States. The commission proposed to the British that they extend an invitation to the FBI and the AEC security staff for a third investigative trip to London. Security standards were still of paramount importance to the Americans. By the end of the year that invitation still had not been extended. The attention of the British was focused on other areas. They had dismissed collaboration for the present and were concentrating their efforts on Hurricane.

Churchill had come to believe that atomic cooperation would be possible only after the British had demonstrated to the Americans that they had something to offer in the atomic exchanges. The British Chiefs of Staff also recognized that until Britain developed an atomic bomb she would not be entitled to play a role in formulating strategies for any future war and, once that war was concluded, in drawing up peace terms. A Chiefs of Staff report of mid-

1952 stated: "We feel that to have no share in what is recognized as the main deterrent in the cold war and the only Allied offensive in a world war would seriously weaken British influence on United States policy and planning in the cold war, and in war would mean that the United Kingdom would have no claim to any share in the policy or planning of the offensive."[27] The detonation of an atomic bomb would underline the United Kingdom's value as a worthy ally, increase her prestige, and win some influence in Washington. The atomic test at Monte Bello became the main focus of the government, since its success could provide the key to the reopening of atomic collaboration with the Americans.

Meanwhile, in preparation for the post-Hurricane negotiations, Churchill appointed a new ambassador to Washington to replace Sir Oliver Franks. He sent over Sir Roger Makins, a man who had been involved in the atomic program from the very beginning. Arneson referred to him as Churchill's "trump card," a "pretty tough customer" who would try to "recapture for the United Kingdom some, if not all, of the role it played with the United States during the past war in the atomic energy field." Nevertheless, Arneson was not impressed; he felt that "the 'trump card' does not look nearly so imposing to us as it undoubtedly must to the British."[28] Arneson knew that no one man had the wherewithal to break down the formidable barriers that the State Department and the AEC had been bumping against for years. In the short term, the results of Makins's efforts would be no better or worse than those of his predecessors.

On October 3, 1952, the British finally detonated its home-grown atomic bomb. The United States had provided not one iota of technical information to the success of the British project. Whereas the British could boast that they had made a major contribution to the successful development of the American atomic bomb, the Americans could make no such claim to the British effort. The British were quite pleased with their achievement, and they had every reason to be. Against all the odds, financial and lack of American cooperation among them, they had achieved an atomic milestone that restored British prestige to the level of a great power. They were one of only three international powers that possessed that new emblem of military superiority, the atomic bomb. Those who had written off the United Kingdom as a waning second-rate power were forced to reassess their judgment. In an address to Parliament, Churchill was magnanimous with his praise: "All those concerned in the production of the first British atomic bomb are to be warmly congratulated on the successful outcome of an historic episode, and I should no doubt pay my compliments to the Leader of the Opposition and the party opposite for initiating it."[29] Dr. William Penney, who had contributed more to the American project than any other British scientist and whose experience at Los Alamos was brought to bear upon the British project, was knighted immediately after the test.

Yet even in the glow of their success, the British tended to view Monte Bello as a means to an end: they still perceived full Anglo-American cooperation as the ultimate goal. Churchill did not doubt that Monte Bello would "lead to a much closer American interchange of information." There were a large number of people in the United States, he explained, who favored closer atomic collaboration, and "this event will greatly facilitate and support the task which these gentlemen have set themselves."[30] But this was not to be. In the last days of the Truman administration, and exactly four weeks after Monte Bello, the Americans exploded a thermonuclear device in the Pacific. With that one explosion they again leaped far ahead of the British in the nuclear field. The British might have found the key to the atomic secret, but now they had to begin the search for that of the hydrogen bomb, and the Americans had no intention of assisting them. An Associated Press survey of incoming congressmen revealed that by a 2 to 1 margin members were against an open exchange of information with the British. As Representative William Harrison of Wyoming saw it, "We would be trading a horse for a rabbit."[31] Cockcroft himself, like others in the British atomic program, admitted that the success of the U.S. hydrogen bomb would make collaboration less likely.[32] Ten months later, in August 1953, the British received another shock. The Soviets detonated a hydrogen bomb. Once again the British were forced to play catch-up.

Churchill still held fast to his optimism that the new U.S. administration under President Dwight D. Eisenhower, who took office in January 1953, would improve things. But that improvement was slow in coming. Despite eleven years of atomic association and even more of international cooperation, full Anglo-American atomic collaboration was still six years in the future. Nevertheless, 1952 had ushered in certain developments that made the collaboration effort seem less critical. The United States' uranium shortage problem was not as acute as in previous years. There was no need for U.S. officials to court the British to gain uranium. On the other hand, the United Kingdom had detonated an atomic bomb. Its scientists now possessed the technical know-how to construct a bomb. The bomb might be relatively primitive, but it provided the British with some security and a great deal of national pride. Attlee and Truman, the men who had inherited the secret wartime agreements and who had attempted to work within and around them, would both be out of office by January 1953. They left Anglo-American atomic cooperation in basically the same state they had found it: a state of nonexistence. Try as they might, both men and their atomic advisers could not resolve the many problems involved. Not until July 3, 1958, did Congress amend the Atomic Energy Act. The amendment gave U.S. presidents the legal authority to exchange information on atomic and hydrogen bombs. It introduced also full Anglo-American atomic cooperation.

Epilogue

The course of Anglo-American atomic energy cooperation was long and uncertain. Eleven years after their atomic association began in 1941, the British and Americans were no closer to creating a full and equal partnership. Nonetheless, the two main problems that had lent a sense of urgency to the postwar collaboration effort resolved themselves. The United States was fast becoming self-sufficient in the area of uranium acquisition, and the United Kingdom had accumulated enough technical expertise to develop an atomic bomb. Both nations, it appeared, had the freedom to reappraise the overall problems and introduce new solutions. But the perceived threat of the worldwide spread of communism injected another intense phase of security concerns into the relationship. What followed over the next six years were more anxieties and frustrations as both nations sought to address the Soviet threat and to overcome the resistance of American opponents to atomic collaboration. The first part of this chapter will summarize those years. The second part will address some of the findings of the preceding study.

When Churchill regained office in 1951, he became acutely sensitive to the financial demands of nuclear research. The result of the U.S. presidential election of November 5, 1952, pleased him. The American people elected Dwight D. Eisenhower as their new president. The president had long opposed the McMahon Act on the grounds that it unfairly and unwisely undermined U.S. relations with its allies. Churchill hoped that the president would help him resolve the financial demands of the British program by providing the United Kingdom with bombs and material. So confident was Churchill that relations would improve that he informed Parliament that he anticipated an atomic partnership that would include both atomic and hydrogen information. Privately, he shared his optimism that, at the least, the United States would turn over atomic bombs to the United Kingdom. Since the arrival of American bombs was imminent, he rejected Lord Cherwell's request to construct another atomic pile to produce more plutonium to feed the United Kingdom's bomb program.[1]

As it turned out, Churchill's optimism was justified. Eisenhower openly challenged the Joint Chiefs of Staff, Lewis Strauss, his choice for the chairmanship of the Atomic Energy Commission (AEC), and the Joint Congressional Committee on Atomic Energy (JCAE) over their restrictive interpretation of the act. However, like Truman before him, he discovered that knocking down the walls of opposition was no easy proposition. The military and Congress still distrusted British security. Moreover, the JCAE believed it had a new secret to protect in the hydrogen bomb. Both Strauss and Secretary of State John Foster Dulles attempted to restrain the president's enthusiasm for collaboration. But he would not be deterred. The president was quick to appreciate that any effective security relationship with the North Atlantic Treaty Organization (NATO) necessitated some sort of nuclear partnership, not just with the United Kingdom but with all the European allies. This was one area where he departed from the previous administration, which had sought a bilateral arrangement with the United Kingdom. On October 30, 1953, he approved NSC 162/2 as a new statement of U.S. national policy and strategy to meet the Soviet threat. The strategy included the following: (1) The United States would rely on its nuclear deterrent and strategic air power; (2) The United States would court foreign powers to gain access to their air bases; (3) The United States would get its allies to assume a greater role in a conventional war, and (4) the United States would push for the rearmament of Germany and Japan. Five weeks later, he adopted NSC 151/2, which proposed sharing with the allies atomic information, atomic weapons, and the means of delivering the weapons. In effect, NSC 151/2 proposed to give the allies some input into NATO atomic planning. But first, he had to win congressional authorization.

Meanwhile, Churchill kept up his demands for a special Anglo-American atomic relationship outside of the NATO umbrella. At separate meetings with the president in Bermuda in December 1953 and in Washington in June 1954, he stated the United Kingdom's position. On August 12, 1953, the Soviets had detonated their own thermonuclear device, thereby placing the United Kingdom at a greater risk of nuclear destruction. Churchill argued that since the United Kingdom did not possess a stockpile of atomic weapons, he saw the need for the storage of a certain number of U.S. atomic bombs in Britain under U.S. custody. Eisenhower was noncommittal at Bermuda, but in Washington he agreed to provide the United Kingdom with warheads in the event of an emergency. By then, the United Kingdom had convinced the Americans that they had taken steps to improve their security system.

In response to Eisenhower's request, the AEC and the Departments of Defense and State had earlier made representations to Congress for an amendment of the McMahon Act. Since uranium self-sufficiency was imminent (new sources had been discovered in Nevada, Wyoming, and Idaho), the administration could no longer effectively use the raw materials argument to justify atomic collaboration. The new argument centered on the theme of common defense and security against Soviet expansionism. And on June 8, 1953, the president made his own public appeal to amend the act.

On August 30, 1954, Congress amended the McMahon Act. However, the legislation fell short of British expectations. It provided for no transfer of information relating to the design and manufacture of nuclear weapons. But it did permit the United States to cooperate with its allies, not just the United Kingdom, on the use of and defense against atomic weapons. In addition, the United States could exchange nonmilitary nuclear technology with its allies. Subsequent meetings with the AEC and Department of Defense in October confirmed the United Kingdom's fears that the amendment had not addressed its needs. The United Kingdom still had to spend time, money, and resources to duplicate atomic and hydrogen information that the United States already had.

Churchill retired on April 5, 1955, to be succeeded by Anthony Eden. By then it was clear to him that the British might have to continue pursuing an independent atomic policy. In late 1954 and early 1955, he reexamined his government's atomic policy and decided to expand the United Kingdom's atomic program. Among other things, the atomic budget was increased to produce more and better bombs and to construct six additional reactors for the production of plutonium and highly enriched uranium. Lord Cherwell had been pressing him to do just that for the past two years. But the prime minister had deluded himself into believing that his personal effort would resurrect the Quebec Agreement. He did succeed in persuading Eisenhower to agree to his publication of the agreement in April 1954. In February, his government also made public its heretofore secret six-month decision to begin research into the development of the hydrogen bomb. This decision was based on the security needs of the United Kingdom and also on the prevailing view among cabinet members that the United Kingdom had once again to prove its worth to the United States. The hydrogen program came to fruition in May 1957, when the British tested a hydrogen bomb. It occurred at the right moment, as events proved.

Between 1955 and 1958, it became clear to the American guardians of nuclear information that the United States possessed little information that the Soviets did not already have. In late August 1957, the Soviet Union successfully tested an intercontinental ballistic missile, thereby placing distant targets within reach of its hydrogen bombs. Moreover, the Soviets also demonstrated that in the sphere of space technology their program surpassed that of the

United States. On October 4 of the same year they launched their Sputnik satellite. These achievements were shocking to many Americans, who, despite the Kremlin's territorial and ideological achievements in Eastern Europe and Asia, had never questioned the United States' technological supremacy. As James R. Killian, Eisenhower's special assistant for science and technology, described this response, "Overnight there developed a widespread fear that the country lay at the mercy of the Russian military machine and that our government and its military arm had abruptly lost the power to defend the homeland itself."[2]

The British and American administrations seized the moment and exploited this fear. On October 10, the new British prime minister, Harold Macmillan, wrote Eisenhower suggesting a pooling of nuclear resources. He received an immediate and friendly response.[3] A summit followed quickly on October 24 and 25. At the summit the president intimated his willingness to seek an amendment to the Atomic Energy Act of 1954. Over the next several months, the administration lobbied Congress on the security benefits of collaborating with the NATO allies in the field of nuclear energy. The menace of the Soviet threat, it had become clear to all, required the concerted effort of the Western allies. The fact that the British had joined the hydrogen club in May 1957 strengthened the argument against secrecy. There was nothing much left to hide. In addition, the British were able to make the compelling argument that finally they had something to contribute to the nuclear deterrence.

On June 19, 1958, Congress passed the amendment to the Atomic Energy Act of 1954. The act ushered in full Anglo-American nuclear exchanges in all areas, including the design, development, and manufacturing of nuclear weapons.

This study has found that Roosevelt and Churchill laid the foundation for the collaboration controversy in the postwar era due to the president's rather ambivalent atomic policy and due to the cavalier manner in which both leaders approached the postwar handling of atomic collaboration. During the early phases of the U.S. atomic program, the president never did dissuade his subordinates in certain terms from promoting the policy of noncooperation. He led them to believe that he supported their recommendation for an independent American program. But in mid-1943, despite the opposition of these atomic energy officials, he stepped forward and arbitrarily imposed his will for closer atomic ties upon them. There is no evidence that the president changed his policy because he disagreed with Bush's and Conant's presentation that an independent program best served the interest of the United States in the present war.[4] He was influenced in his decision by the exigencies of the war, Churchill's persistent request for a partnership, and his close personal relationship with Churchill. It is equally doubtful that the president foresaw the bomb in the context of postwar international relations. He had not carefully considered the implications of atomic weapons in a postwar world. Undoubtedly, Churchill, thanks to his advisers, had given the matter some thought. In fact, as Bush himself later confessed, the president's atomic advisers had not advised him fully on the postwar ramifications of atomic power. In the latter stages of the war, Churchill persuaded Roosevelt to sign an agreement to continue the atomic partnership in the postwar era. For reasons unknown, the president did not disclose the existence of this agreement to his advisers. It is disputable that the president had bought into Churchill's argument to create a postwar atomic partnership to promote peace in Europe or that he had any plans of his own to use his atomic partner as a sort of peace surrogate in Europe. What we do know is that this Hyde Park Aide-Memoire became the foundation upon which the postwar collaboration controversy was built.

The United States and the United Kingdom continued their partnership after 1945. The Soviet Union and its communist ideology seemed to threaten the values of Western capitalism and the democratic freedom of the liberated peoples of Eastern Europe. However, this U.K.-U.S. special relationship did not extend into the area of atomic energy. Nor was this a relationship between equal partners. The United States had emerged from the war as the leader of the free world, while the United Kingdom came away as a power in distress. The war had wrecked the British economy, forcing the Labour government to pursue a tentative foreign policy. Both within and without the realm of atomic matters, the United States viewed the United Kingdom as a junior partner, and the United Kingdom conducted itself as such. It had become a subservient power that looked to the United States to bail it out of national and international situations. Fortunately, from the standpoint of British prestige, U.S. officials allowed it to hold on to the United States' coattails as both nations battled the spread of communism. But it was difficult for congressional leaders and bureaucrats to accept the United Kingdom as a power worthy of sharing U.S. atomic secrets.

As the debate unfolded within the various departments, a common theme emerged. It concerned the perils and political uses of bureaucracy within and among modern states and the seemingly inevitable victory of realpolitik and old national loyalties over international ideals in the atomic age. The interest of the national state was always paramount. Atomic bombs were unlike any weapon controlled previously by any nation. The use of a single instrument held the potential to determine the outcome of a conflict. There was reluctance on the part of some U.S. government officials to share this secret with any other power, friend or foe. Atomic weapons afforded the United States the opportunity to leap to the pinnacle of world power. The United States could not pass it up. Less than 200 years after its creation and 75 years after a divisive civil war, the nation was poised to take its place as the most powerful military and economic entity in the world. But this supremacy could be guaranteed, in the minds of these officials, only if the United States maintained its monopoly of atomic weapons. Although it was clear to those in the know that the "secret" could not be guarded in perpetuity, they still attempted to postpone and control the inevitable. For the most part, they believed that since the "secret" could not be protected indefinitely, its awesome power should be entrusted to an international agency controlled by the United States or should remain in the trust of the United States itself, a peace-loving country with no expansionist ambitions.

The congressional "watchdogs" on the JCAE knew that the public supported their mission to prevent the proliferation of atomic weapons. After all, the United States owned the patent on bomb development. These men's views mirrored those of a public caught up both in the euphoria of patriotism fed by a newfound U.S. international power and also in the growing realization of the dangers of nuclear warfare. The United States had emerged from the war victorious, powerful, and in possession of the bomb. Americans believed that this international prestige and atomic power could be used to maintain global peace and prevent future wars. This was no time to share the secret with individual powers. A large sector of the public also bought into the idealistic argument that proliferation might be monitored through the policing efforts of the United Nations. The atomic arguments waged in the media centered not on the merits of bilateral atomic arrangements but rather on the issue of national or international control of atomic energy.

It is indisputable that uranium shortages in the United States played a pivotal role in influencing the on-again, off-again collaboration policy of the United States. The JCAE's solution to the shortages was to force the United Kingdom out of the bomb-making business. The United States could accumulate a stockpile of bombs to create its Pax Americana only if it gained access to all available sources of uranium. To achieve this exclusive control, it had to impede the United Kingdom's attempts to engage in bomb production, thereby minimizing uranium competition. The JCAE urged the administration to pressure the United Kingdom into accepting the U.S. monopoly of atomic weapons, especially since those weapons were earmarked for the general security of both nations. The administration, however, mindful of its international image, wooed the United Kingdom with the promise of atomic information in exchange for British uranium. This atomic policy prolonged the agony of noncollaboration while it kept up production in the Americans' atomic plants.

It appears that whenever British officials were on the verge of pursuing an independent course, their American counterparts deterred them by reviving the idea of renewed negotiations. Understandably, since their program could have benefited, the British jumped at the prospect of collaboration. The result ended always in painful frustrations for the United Kingdom and more uranium for the United States.

Undoubtedly, the administration would have instituted full atomic collaboration with the British. It made promises on and off the record that it was quite prepared to honor. This was true in early 1946, when Truman backed out of his agreement to continue the wartime atomic relationship, thanks to Groves. It was evidenced again when the Modus Vivendi was negotiated. The administration and the AEC were willing to concede to the United Kingdom an atomic bomb program and had even initiated steps to share bomb information. But they encountered opposition from AEC Commissioner Lewis Strauss and Senators Hickenlooper and Vandenberg. Their opposition led to a breakdown of the Modus. It also contributed to U.K.-U.S. disagreements over what actually had been promised. As it had done in 1946, the administration hid behind outright denials and semantics, rather than acknowledging its explicit and implicit commitments and admitting that it was unable to honor them due to the role of the "watchdogs." This duplicity led its U.K. counterpart to the incorrect conclusion that the United States, from the outset, had no intention of keeping its commitments.

Not until 1949 did the administration succeed in convincing congressional leaders of the security benefits of cooperating with the British. This was achieved only after Truman threatened a constitutional confrontation to implement his atomic policy. He felt that the security of the United States, as provided for by the McMahon Act, warranted a policy of Anglo-American atomic cooperation. Moreover, the constitution gave him, not Congress, the authority to formulate foreign policy. The integration proposal grew out of the compromise between the president and Congress.

The British decision to even consider the integration proposal was not an easy one. From the inception of atomic weapons research, politicians, scientists, and bureaucrats on both sides of the Atlantic recognized the potential diplomatic and military importance of atomic weapons. No one was positioned to articulate clearly how the new weapon would change the international hierarchy, but both British and American officials predicted superpower status for the possessor of atomic weapons. The destruction of Hiroshima and Nagasaki surpassed even the wildest exaggeration of those who dared describe the power of atomic weapons. The United States, the United Kingdom, the USSR, and France were driven to harness their own resources, and those of other countries, to master the new technology at all costs. The British made enormous social and financial sacrifices to invest in an atomic program. At a time of great postwar economic distress, they secretly allocated large sums to the project without attracting criticism. Whereas the American postwar atomic program was conducted in a relatively open political climate, where political leaders were accountable to the public and there was a raging atomic debate in and out of the government, the program of the British was conducted with the utmost secrecy. At a time when the United Kingdom, primarily for financial reasons, was withdrawing from engagements and commitments around the world, exorbitant sums were pumped into atomic research. Revealingly, two economically disadvantaged nations, the United Kingdom and the Soviet Union, sought to manufacture a technologically advanced weapon that neither was financially equipped to support without creating hardships for its people. The Soviet economy never did expand to the point where it could sustain both a nuclear program and the basic economic needs of its people. Ultimately, its failure to do so contributed to the demise of the Soviet system in 1991. But the realities of the atomic age, national pride, power politics, and perceived security concerns necessitated the acquisition of atomic weapons, regardless of the sacrifices involved. Neither the United Kingdom nor the Soviet Union could risk conceding an atomic monopoly to the United States.

Ironically, the failure to achieve Anglo-American atomic collaboration strengthened the security bond existing between both countries. Duplicating time-consuming experiments already concluded in the United States delayed the development of a British bomb. Hence, in the face of Soviet expansionism and in the shocking awakening of the Soviet detonation of 1949, the United Kingdom, since it had no atomic weapons in hand, had little choice but to place its security in the hands of the United States. Despite its concerns and sacrifices, the British government had to seriously consider accepting the American integration proposal. Military and international realities shot down the ideals of atomic nationalism. An independent deterrence program with no bombs in hand afforded no security against a Soviet Union with the means to stockpile atomic weapons and deliver them into the heart of Britain.

Not even the first two spy scandals could derail this integration effort. In fact, the spy scandals made the need for integration even more urgent because neither side had any way of telling how much the Soviets had benefited from Klaus and Pontecorvo. It might take a combined effort to remain ahead of them. Had the third spy scandal not broken, the United States and the United Kingdom would have integrated their atomic program, thereby ushering in full atomic collaboration. By defecting to the Soviet Union, Maclean and Burgess did one last service for their handlers by destroying, for the foreseeable future, the idea of Anglo-American atomic collaboration. Notwithstanding the breakdown, however, the alliance remained intact and firmly committed to the worldwide destruction of communist ideologies by means of U.S. military power.

Congressional opposition to any sort of atomic collaboration remained constant throughout the atomic debate. Over the years, the Truman and Eisenhower administrations chipped away at the restrictive features of the Atomic Energy Act. But in the final analysis, Congress made concessions only after international atomic developments made it clear that there was no secret, atomic or hydrogen, to protect. By 1958, both the USSR and the United Kingdom had improved their military technology to the point where they were able to share the thermonuclear stage with the United States. In addition, the Soviet system and the many international crises it engendered seemed to threaten the security of the free world and that of the United States. Congress finally accepted the wisdom of sharing all nuclear information with its allies. It would take a global alliance, built on mutual trust, to contain the global threat of communism. As always, the decision was based on protecting the security interest and ideals of the United States.

Notes

CHAPTER ONE

1. Richard G. Hewlett and Oscar E. Anderson, *The New World, 1939–1946,* vol. 1 of *A History of the United States Atomic Energy Commission* (Pittsburgh: Pennsylvania State University Press, 1962), 12–13.

2. Bertrand Goldschmidt, *Atomic Rivals,* trans. Georges M. Temmer (New Brunswick, N.J.: Rutgers University Press, 1990), 48.

3. Margaret Gowing, *Britain and Atomic Energy, 1939–1945* (New York: St. Martin's Press, 1964), 28.

4. Ibid., 40.

5. Ibid., 34.

6. Quoted in Richard Rhodes, *The Making of the Atomic Bomb* (New York: Simon and Schuster, 1986), 292.

7. Ibid., 292–95; also Hewlett and Anderson, *New World,* 15.

8. Rhodes, *Making of the Atomic Bomb,* 293.

9. Goldschmidt, *Atomic Rivals,* 52; Gowing, *Britain and Atomic Energy,* 35.

10. Goldschmidt, *Atomic Rivals,* 52.

11. Ibid.

12. Gowing, *Britain and Atomic Energy,* 36.

13. Quoted in ibid.

14. Ibid., 37.

15. Ibid.

16. Peter Pringle and James Spigelman, *The Nuclear Barons* (New York: Holt, Rinehart and Winston, 1981), 81.

17. Ibid., 36.

18. The following information on Joliot and his colleagues can be found in Goldschmidt, *Atomic Rivals,* 50–54, 74–80.

19. Ibid.

20. Ibid.

21. Quoted in Rhodes, *Making of the Atomic Bomb,* 303.

22. Ibid., 314.

23. Ibid., 317.

24. Gowing, *Britain and Atomic Energy,* 42.

25. A detailed explanation of this study can be found in Gowing's *Britain and Atomic Energy,* 45–89.

26. Quoted in Gowing, *Britain and Atomic Energy,* 53.

27. Quoted in Rhodes, *Making of the Atomic Bomb,* 337.

28. Ibid.

29. Ibid., 360; also Hewlett and Anderson, *New World,* 35–36.

30. Quoted in Rhodes, *Making of the Atomic Bomb,* 362.

31. Hewlett and Anderson, *New World,* 38–39; also Rhodes, *Making of the Atomic Bomb,* 368.

32. Quote is from page 394 of the report. For the full text of the report, see Appendix 2 in Gowing's *Britain and Atomic Energy,* pp. 394–436.

33. "Lauritsen to Bush," July 11, 1941, Vannevar Bush–James B. Conant Files, Office of Scientific Research and Development (Record Group 227), National Archives, Washington, D.C. (hereafter cited as Bush-Conant Files), fol. 9.

34. Quoted in Rhodes, *Making of the Atomic Bomb,* 372.

35. Martin J. Sherwin, *A World Destroyed: The Atomic Bomb and the Grand Alliance* (New York: Alfred A. Knopf, 1975), 36.

36. "Lauritsen to Bush," July 11, 1941, Bush-Conant Files, fol. 9.

37. "Cherwell to Prime Minister," August 27, 1941, War of 1939–1945: Correspondence and Papers (AB 1), Public Record Office, London (hereafter cited as PRO, AB 1), PRO, AB 1/170.

38. "Prime Minister to General Ismay," August 30, 1941, Operations Papers (PREM 3), Public Record Office, London (hereafter cited as PRO, PREM 3), PREM 3/139/8A.

39. Ibid.

40. James G. Hershberg, *James B. Conant: Harvard to Hiroshima and the Making of the Nuclear Age* (New York: Alfred A. Knopf, 1993), 149–50.

41. "President to Prime Minister," October 11, 1941, PRO, AB 1/207.

42. "Norman Brooks to Martin," December 10, 1941, PRO, PREM 3/139/8A.

43. "Prime Minister to President," December [no day given], 1941, PRO, AB 1/207.

44. The *Prince of Wales* was one of the vessels that had sunk the *Bismarck.* Four months after Placentia, the Japanese sank the *Prince of Wales* in the Pacific, resulting in the loss of most of the crew present at this summit.

45. Quoted in Herbert Feis, *Churchill, Roosevelt, Stalin: The War They Waged and the Peace They Sought* (Princeton, N.J.: Princeton University Press, 1957), 7.

46. "Report to the National Academy of Science by the Committee on Uranium," November 6, 1941, Bush-Conant Files, fol. 18.

47. Quoted in "Summary," President's Secretary's Files, Historical Files, Truman Library, Independence, Mo., box 227.

48. "Bush to Darwin," December 23, 1941, PRO, AB 1/602.

49. Gowing, *Britain and Atomic Energy,* 119.

50. "Report of Meeting," November 27, 1941, PRO, AB 1/207.

51. "Bush to Darwin," December 23, 1941, PRO, AB 1/207.

52. "Brooks to Hovde for Carol Wilson and Dr Bush," Bush-Conant Files, fol. 8; see also PRO, AB 1/207.

53. Gowing, *Britain and Atomic Energy,* 126.

54. "Akers to Anderson," February 24, 1942, PRO, AB 1/34.

55. Gowing, *Britain and Atomic Energy,* 132.

56. "Anderson to Bush," March 23, 1942, Bush-Conant Files, fol. 9.

57. "Bush to Anderson," April 20, 1942, Bush-Conant Files, fol. 9.

58. Quoted in Gowing, *Britain and Atomic Energy,* 139.

59. "Anderson to Prime Minister," July 30, 1942, PRO, PREM 3/139/8A.

60. Ibid.

61. Goldschmidt, *Atomic Rivals,* 133.

62. Ibid., 134.

63. Winston Churchill, *The Hinge of Fate,* vol. 4 of *The Second World War* (Boston: Houghton Mifflin, 1950), 379–81.

CHAPTER TWO

1. "Report to Akers and Halban," January 22, 1943, War of 1939–45: Correspondence and Papers (AB 1), Public Record Office, London (hereafter cited as PRO, AB 1), AB 1/128.

2. "Halban to Akers," May 13, 1942, PRO, AB 1/357.

3. "Perrin to Akers," June 8, 1942, PRO, AB 1/357.

4. "Anderson to Bush," August 5, 1942, Vannevar Bush–James B. Conant Files, Office of Scientific Research and Development (Record Group 227), National Archives, Washington, D.C. (hereafter cited as Bush-Conant Files), fol. 9; see also Harrison-Bundy Files, Records of the Office of the Chief of Engineers, Manhattan Engineer District (Record Group 77), National Archives, Washington, D.C. (hereafter cited as Harrison-Bundy Files), file 47.

5. "Anderson to Bush," August 5, 1942, Bush-Conant Files, fol. 9.

6. "Bush to Anderson," September 1, 1942, Bush-Conant Files, fol. 9; see also Harrison-Bundy Files, file 47.

7. "Conant to Bush," n.d., Bush-Conant Files, fol. 9.

8. "Report by Akers and Halban," January 22, 1943, PRO, AB 1/128.

9. "Conant to Bush," October 26, 1942, Bush-Conant Files, fol. 16.

10. "Excerpt from the Diary of S/W [secretary of war]," October 29, 1942, Records of the Special Assistant to the Secretary of State for Atomic Energy Matters, Lot 57 D688, General Records of the Department of State, National Archives, Suitland Branch, Md. (hereafter cited as General Records of the Department of State), box 56.

11. Up until January 12, 1943, the only documented correspondence between

Churchill and Roosevelt on atomic energy was the initial letter of October 1941 from the president seeking collaboration and the responding letter of December 1941 from the prime minister expressing readiness to collaborate.

12. "Akers to Perrin," November 5, 1942, PRO, AB 1/357.

13. "Akers to Perrin," November 16, 1942, PRO, AB 1/128.

14. Ibid.

15. "Bush to Military Policy Committee," December 15, 1942, Bush-Conant Files, fol. 4; see also Harrison-Bundy Files, file 47.

16. "Akers to Conant," December 15, 1942, Bush-Conant Files, fol. 9; see also PRO AB 1/128 and General Records of the Department of State, box 56.

17. "Bush to Military Policy Committee," December 15, 1942, Bush-Conant Files, fol. 4.

18. Ibid.

19. Ibid.

20. Margaret Gowing, *Britain and Atomic Energy, 1939–1945* (New York: St. Martin's Press, 1964), 155. This point of view is also expressed in Richard G. Hewlett and Oscar E. Anderson, *The New World, 1939–1946*, vol. 1 of *A History of the United States Atomic Energy Commission* (Pittsburgh: Pennsylvania State University Press, 1962), 267–68.

21. Martin J. Sherwin, *A World Destroyed: The Atomic Bomb and the Grand Alliance* (New York: Alfred A. Knopf, 1975), 172.

22. James G. Hershberg, *James B. Conant: Harvard to Hiroshima and the Making of the Nuclear Age* (New York: Alfred A. Knopf, 1993), 183.

23. "Anderson to Prime Minister," January 11, 1943, Operations Papers (PREM 3), Public Record Office, London (hereafter cited as PRO, PREM 3), PREM 3/139/8A.

24. "Conant to Dean Mackenzie," January 2, 1943, PRO, PREM 3/139/8A.

25. "Canada (HC) to Dominion Office," January 8, 1943, PRO, AB 1/374.

26. "Akers to Gordon Munro," January 30, 1943, PRO, AB 1/128. (In this letter, Akers informed Munro about his comments to Mackenzie.)

27. Hershberg states that eight months later Akers finally concluded that the British officials should cultivate "really good relations" with Conant, who strongly influenced the decisions of Groves and Bush. Hershberg, *James B. Conant*, 181.

28. Quoted in Hershberg, *James B. Conant*, 160.

29. Quoted in Sherwin, *A World Destroyed*, 47.

30. Hershberg, *James B. Conant*, 164.

31. "Akers to Gordon Munro," January 18, 1943, PRO, AB 1/374.

32. "Canada (HC) to Dominion Office," January 21, 1943, PRO, AB 1/374.

33. "Conant to Akers," December 15, 1942, PRO, AB 1/357.

34. "Akers to Perrin," December 29, 1942, PRO, AB 1/357.

35. "Anderson to Prime Minister," January 11, 1943, PRO, PREM 3/139/8A.

36. "Notes on Discussions with Groves, Conant and Akers," January 26, 1943, PRO, AB 1/128.

37. "Perrin to Akers," January 19, 1943, PRO, AB 1/374.

38. "Perrin to Akers," January 20, 1943, PRO, AB 1/374.

39. "Perrin to Tube Alloys Committee," January 27, 1943, PRO, AB 1/374.

40. The correspondences between Churchill and Hopkins can be found in PRO, PREM 3/139/8A.

41. Ibid.

42. Ibid.

43. "Dominion Office to Anderson," February 27, 1943, PRO, AB 1/374.

44. "Anderson to Dominion Office," March 1, 1943, PRO, AB 1/374.

45. "Anderson to Dominion Office," March 3, 1943, PRO, AB 1/374.

46. Hershberg, *James B. Conant,* 186–87.

47. "Bush to Hopkins," March 31, 1943, Bush-Conant Files, fol. 10.

48. "Conant to Bush," March 25, 1943, Bush-Conant Files, fol. 10. Also in Hershberg, *James B. Conant,* 187, and Sherwin, *A World Destroyed,* 80–81.

49. Ibid.

50. Ibid.

51. Ibid.

52. Guy Hartcup and T. E. Allibone, *Cockcroft and the Atom* (Bristol, England: Adam Hilger, 1984), 125.

53. "Prime Minister to Hopkins," April 1, 1943, PRO, PREM 3/139/8A.

54. "Cooperation with the USA," April 1, 1943, PRO, AB 1/374.

55. "Prime Minister to Lord President of the Council," April 15, 1943, PRO, PREM 3/139/8A.

56. "High Commissioner to Canada to the Lord President," July 5, 1943, PRO, AB 1/376.

57. "Note to Talk with Dean Mackenzie," May 14, 1943, PRO, AB 1/374.

58. "Lord President to Prime Minister," May 15, 1943, PRO, PREM 3/139/8A.

59. "Memo of Conversation," May 25, 1943, Bush-Conant Files, fol. 10.

60. "Prime Minister to Lord President," May 26, 1943, Churchill Papers, Chartwell Trust, Churchill College, Cambridge, England, 20/128; see also PREM 3/139/8A.

61. "Memo of Conversation with the President," June 24, 1943, Bush-Conant Files, fol. 10.

62. "Memo of Conversation," May 25, 1943, Bush-Conant Files, fol. 10.

63. "Memo of Conversation with the President," June 24, 1943, Bush-Conant Files, fol. 10.

64. "Prime Minister to Hopkins," July 9, 1943, PRO, PREM 3/139/8A.

65. Warren F. Kimball, ed., *Alliance Forged: November 1942–February 1944,* vol. 2 of *Churchill and Roosevelt: The Complete Correspondence* (Princeton, N.J.: Princeton University Press, 1984), 351.

66. "President to Bush," July 20, 1943, Bush-Conant Files, fol. 10.

67. "Lord President to Canadian Embassy," July 2, 1943, PRO, AB 1/376.

68. "Talk with Dr Bush," July 15, 1943, PRO, AB 1/376.

69. "Memo to Dr Conant," July 23, 1943, Harrison-Bundy Files, file 47; also in Bush-Conant Files, fol. 10.

70. Ibid.

71. Ibid.

72. "Prime Minister to Lord President of the Council," July 18, 1943, PRO, PREM 3/139/8A.

73. "Memo to Conant," July 23, 1943, Bush-Conant Files, fol. 10.

74. "Prime Minister to Lord President of the Council," July 18, 1943, PRO, PREM 3/139/8A.

75. "Akers to Perrin," August 19, 1943, PRO, AB 1/376.

76. Ibid.

77. "Perrin to Akers," August 11, 1943, PRO, AB 1/376.

78. "Memorandum of Meeting at 10 Downing St. on July 22, 1943," General Records of the Department of State, box 56; see also Harrison-Bundy Files, file 47.

79. Ibid.

80. "Memo to Dr Conant," July 23, 1943, Bush-Conant Files, fol. 10.

81. "Memo for the File," August 4, 1943, Bush-Conant Files, fol. 10. It turned out that during transmission to Bush the key word *renew* somehow was changed to *review.*

82. "Conant to Bush," July 30, 1943, General Records of the Department of State, box 56.

83. "Conant to Bush," August 3, 1943, Bush-Conant Files, fol. 9.

84. "Conant to Bush," August 6, 1943, Bush-Conant Files, fol. 9.

85. Ibid.

86. "Anderson to Prime Minister," July 23, 1943, and "Akers to Lord President," July 23, 1943, PRO, PREM 3/139/8A.

87. "Anderson to Prime Minister," July 24, 1943, PRO, PREM 3/139/8A.

88. Copies of the agreement can be found in General Records of Department of State, box 56; Harrison-Bundy File, fol. 49; and the appendix of Gowing's *Britain and Atomic Energy,* 439.

89. "Churchill to Mackenzie King," August 11, 1943, PRO, PREM 3/139/8A.

90. Hershberg, *James B. Conant,* 188–89.

91. Sherwin, *A World Destroyed,* 84.

CHAPTER THREE

1. Quoted in Jonathan E. Helmreich, *Gathering Rare Ores: The Diplomacy of Uranium Acquisition, 1943–54* (Princeton, N.J.: Princeton University Press, 1986), 10.

2. U.S. Congress, Joint Congressional Committee on Atomic Energy, "Study of the British Disclosure," February 4, 1950, Classified Transcripts of Meetings Held in Executive Session, Center for Legislative Archives, National Archives,

Washington, D.C. (hereafter referred to as JCAE Classified Transcripts), Appendix IV, box 5, doc. 1371.

3. "Supply of Material," June 11, 1943, Operations Papers (PREM 3), Public Record Office, London (hereafter cited as PRO, PREM 3), PRO, PREM 3/139/8A.

4. John Wheeler-Bennet, *John W. Anderson, Viscount Waverly* (London: Macmillan, 1962), 289.

5. Ibid.

6. Helmreich, *Gathering Rare Ores,* 13.

7. Ibid., 11.

8. Ibid., 10.

9. Ibid., 11; see also "Diplomatic History of the Manhattan Project," Harrison-Bundy Files, Records of the Office of the Chief of Engineers, Manhattan Engineer District (Record Group 77), National Archives, Washington, D.C. (hereafter cited as Harrison-Bundy Files), file 111.

10. U.S. Department of State, "Agreement between the United States and the United Kingdom," in *Foreign Relations of the United States: General; Economic and Social Matters, 1944,* vol. 2 (Washington, D.C.: Government Printing Office, 1967), 1026–28.

11. In 1946, while on a visit to the United States, Edgar Sengier was awarded the Medal of Merit for meritorious services to the United States in the area of nuclear raw materials. Groves recommended and rushed through its passage in just over a week. It was hoped that the medal would serve as inducement for continued service.

12. "Memo for Dr Conant," August 20, 1943, Vannevar Bush–James B. Conant Files, Office of Scientific Research and Development (Record Group 227), National Archives, Washington, D.C. (hereafter cited as Bush-Conant Files), fol. 9.

13. "Negotiations with the Americans," September 13, 1943, War of 1939–1945: Correspondence and Papers (AB 1), Public Record Office, London (hereafter cited as PRO, AB 1), PRO AB 1/129.

14. "Prime Minister to Anderson," May 27, 1944, PRO, PREM 3/139/11A.

15. "Akers to Perrin," August 31, 1943, PRO, AB 1/376.

16. "Stewart to Chairman and Members of the Executive Committee of S-1," September 10, 1943, Dr. Lyman Briggs Alphabetical Files (Record Group 227), Office of Scientific Research and Development, National Archives, Washington, D.C., FRC, box B, Subject File, "Meeting, Executive Committee" Folder.

17. The agreement can be found in Special Assistant to the Secretary of State for Atomic Energy Matters, Lot 57 D688, General Records of the Department of State, National Archives, Suitland Branch, Md. (hereafter cited as General Records of the Department of State), box 56; also in Harrison-Bundy Files, file 49.

18. "Prime Minister to Lord President," August 25, 1943, Churchill Papers, Chartwell Trust, Churchill College, Cambridge, England, 20/128; see also PRO, PREM 3/139/8A.

19. "Oliphant to Perrin," November 17, 1943, PRO, AB 1/376.

20. "Collaboration with the American Group," October 18, 1943, PRO, AB 1/376.

21. Quoted in Helmreich, *Gathering Rare Ores,* 101.

22. "Webster to Perrin," December 21, 1943, PRO, AB 1/264.

23. "Chadwick to Sir Edward Appleton," January 25, 1945, PRO, AB 1/264.

24. R. Gordon Arneson, interview by Neil M. Johnson, June 21, 1989, transcript, Oral History Collection, Truman Library, Independence, Mo., interview 456.

25. "Chadwick to Anderson," February 22, 1945, Sir R. Campbell and Mr Nevile Butler Papers, Private Collections: Ministers and Officials (FO 800), Public Record Office, London (hereafter cited as PRO, FO 800), FO 800/524.

26. Ibid.

27. Ibid.

28. "Anderson to Chadwick," March 7, 1945, PRO, FO 800/524.

29. Bertrand Goldschmidt, *Atomic Rivals,* trans. Georges M. Temmer (New Brunswick, N.J.: Rutgers University Press, 1990), 203.

30. "Notes on Contact with French on T.A.," n.d., Harrison-Bundy Files, file 36.

31. "Anderson to Bush," August 5, 1942, Bush-Conant Files, fol. 9.

32. "Bush to Anderson," September 1, 1942, Bush-Conant Files, fol. 9.

33. "Memorandum to the Secretary of War," December 23, 1944, Maj. Gen. Leslie R. Groves Files, Records of the Office of the Chief of Engineers, Manhattan Engineer District (Record Group 77), National Archives, Washington, D.C. (hereafter cited as Gen. Groves Files), file 26C.

34. "Notes on Contact with French on T.A.," n.d., Harrison-Bundy Files, file 36.

35. "Memorandum to the Secretary of War," December 23, 1944, Gen. Groves Files, file 26C.

36. "Aide-Memoire," n. d., Harrison-Bundy Files, file 18.

37. "Meeting between the President, Secretary of War and Major Gen. Groves," December 30, 1944, Gen. Groves Files, file 24.

38. "Groves to Secretary of War," December 14, 1944, Harrison-Bundy Files, file 36.

39. "Meeting between the President, Secretary of War and Major Gen. Groves," December 30, 1944, Gen. Groves Files, file 24.

40. "Summary," August 23, 1945, PRO, FO 800/547.

41. "Notes on Contact with French on T.A.," n.d., Harrison-Bundy Files, file 36.

42. "Halifax to Bundy," February 23, 1945, Harrison-Bundy Files, file 36.

43. "General Groves' Own Views on Handling the French," January 20, 1945, Harrison-Bundy Files, file 36.

44. Quoted in Margaret Gowing, *Britain and Atomic Energy, 1939–1945* (New York: St. Martin's Press, 1964), 345.

45. "Churchill Insists on Secrecy," March 25, 1945, Appendix H, in Martin J. Sherwin, *A World Destroyed: Hiroshima and the Origins of the Arms Race* (New York: Vintage Books, 1975), 290–91.

46. "General Groves' Own Views on Handling the French," January 20, 1945, Harrison-Bundy Files, file 36.

47. "From C. D. Howe to CPC," March 6, 1945, Harrison-Bundy Files, file 36.

48. Margaret Gowing, *Policy Making,* vol. 1 of *Independence and Deterrence: Britain and Atomic Energy, 1945–1952* (London: Macmillan, 1974), 10.

49. "Field Marshall Wilson to Chancellor of the Exchequer," May 26, 1945, PRO, FO 800/525.

50. "Cherwell to Anderson," June 30, 1944, PRO, PREM 3/139/11A.

51. Copies of the Hyde Park Aide-Memoire can be found in PREM 3/139/8A and 11A and in General Records of the Department of State, box 56.

52. "Prime Minister to Lord Halifax for Lord Cherwell," September 20, 1944, PRO, PREM 3/139/8A.

53. Sherwin, *A World Destroyed,* 113–14.

54. "Note," October, 1951, President's Secretary's Files, Historical Files, Truman Library, Independence, Mo., box 227.

55. "Memorandum for Dr Conant," September 25, 1944, Bush-Conant Files, fol. 38; James G. Hershberg, *James B. Conant: Harvard to Hiroshima and the Making of the Nuclear Age* (New York: Alfred A. Knopf, 1993), 204; Sherwin, *A World Destroyed,* 122–23.

56. "Memorandum for Dr Conant," September 25, 1944, Bush-Conant Files, fol. 38.

57. "Memorandum to Dr Conant," September 23, 1944, Bush-Conant Files, fol. 19A.

58. U.S. Congress, Records of the Joint Congressional Committee on Atomic Energy, Declassified Materials from Classified Boxes, Series C, General Subject Files, Center for Legislative Archives, National Archives, Washington, D.C., box 22, doc. 2157.

CHAPTER FOUR

1. James F. Byrnes, *Speaking Frankly* (New York: Harper and Brothers Publishers, 1947), 257.

2. "Memorandum for the President," March 3, 1945, Harrison-Bundy Files, Records of the Office of the Chief of Engineers, Manhattan Engineer District (Record Group 77), National Archives, Washington, D.C. (hereafter cited as Harrison-Bundy Files), file 8.

3. "Chancellor of the Exchequer to the Secretary of War," July 18, 1945, Re-

cords of the Special Assistant to the Secretary of State for Atomic Energy Matters, Lot 57 D688, General Records of the Department of State, National Archives, Suitland Branch, Md. (hereafter cited as General Records of the Department of State), box 55.

4. Jonathan E. Helmreich, *Gathering Rare Ores: The Diplomacy of Uranium Acquisition, 1943–54* (Princeton, N.J.: Princeton University Press, 1986), 8.

5. Quoted in Helmreich, *Gathering Rare Ores,* 110.

6. "Truman and the Atom Bomb," Papers of Eben A. Ayers, Subject File, Truman Library, Independence, Mo. (hereafter cited as Papers of Eben A. Ayers), box 4; see also [Untitled summary by Eben A. Ayers], October 1951, President's Secretary's Files, Historical Files, Truman Library, Independence, Mo. (hereafter cited as President's Secretary's Files), box 227.

7. That committee would include Bush, Conant, Byrnes, and Stimson. A scientific panel of Dr. Enrico Fermi, Dr. E. O. Lawrence, Dr. Arthur Compton, and Dr. Robert Oppenheimer assisted it.

8. Henry L. Stimson, "The Henry Lewis Stimson Diaries," Stimson Literary Trust, Manuscripts and Archives, Yale University, New Haven, Conn., 68–72 (copy used was found at Truman Library).

9. Quoted in Martin Gilbert, *Churchill: A Life* (London: William Heinemann, 1991), 598.

10. Margaret Gowing, *Policy Making,* vol. 1 of *Independence and Deterrence: Britain and Atomic Energy, 1945–1952* (London: Macmillan, 1974), 5.

11. "Anderson to Prime Minister," January 19, 1945, Operations Papers (PREM 3), Public Record Office, London (hereafter cited as PRO, PREM 3), PREM 3/139/11A.

12. "Sir Henry Tizard to Prime Minister," April 4, 1945, PRO, PREM 3/139/ 11A.

13. "Prime Minister to General Ismay," April 19, 1945, PRO, PREM 3/139/ 11A.

14. The Labour government updated the report after the bombing of Japan.

15. Hugh Dalton, "Hugh Dalton Diary," July 12, 1945, Hugh Dalton Diaries and Papers, British Library of Economics, London School of Economics, London.

16. "Note of a Meeting of Ministers," August 10, 1945, Ad-Hoc Committees: Gen and Miscellaneous Series (CAB 130), Public Record Office, London (hereafter cited as PRO, CAB 130), CAB 130/2.

17. Alan Bullock, *Ernest Bevin: Foreign Secretary, 1945–51* (London: W. W. Norton, 1983), 185–86.

18. Quoted in Francis Williams, *A Prime Minister Remembers: The War and Post-War Memoirs of the Rt. Hon. Earl Attlee* (London: William Heinemann, 1961), 95.

19. "Prime Minister to the President," August 16, 1945, Papers of Harry S. Truman, Post-Presidential Files, Truman Library, Independence, Mo. (hereafter cited as Post-Presidential Files), box 3.

20. "President Press Conference," October 8, 1945, Papers of Eben A. Ayers, box 4.

21. "President Press Conference," October 31, 1945, Papers of Eben A. Ayers, box 4.

22. From the last months of the war to early 1947, Makins was the official at the British embassy responsible for atomic energy matters. Between 1947 and 1952, he returned to the United Kingdom to serve as assistant and then deputy undersecretary in the Foreign Office. Throughout, he remained in close contact with the atomic program and all its problems.

23. "Meeting of Secretaries of State, War and Navy," October 10, 1945, Maj. Gen. Leslie R. Groves Files, Records of the Office of the Chief of Engineers, Manhattan Engineer District (Record, Group 77), National Archives, Washington, D.C. (hereafter cited as Gen. Groves Files), file 13.

24. "From Washington to Foreign Office," September 25, 1945, Prime Minister's Office: Correspondence and Papers, 1945–51 (PREM 8), Public Record Office, London (hereafter cited as PRO, PREM 8), PRO, PREM 8/117; see also Private Collections: Ministers and Officials: Various (FO 800), Public Record Office, London (hereafter cited as PRO, FO 800), PRO, FO 800/535.

25. "Prime Minister to the President," September 25, 1945, President's Secretary's Files, box 170.

26. Dean Acheson, *Present at the Creation: My Years in the State Department* (New York: W. W. Norton, 1969), 123–24.

27. The diverse views expressed at these meetings will be discussed in chapter 5.

28. "Atomic Energy," October 3, 1945, President's Secretary's Files, box 227.

29. "President to Prime Minister," October 5, 1945, Papers of Harry S. Truman, Post-Presidential Files, box 3.

30. Some scholars have interpreted Attlee's follow-up query and the president's seemingly slow and reluctant reply of October 13 as an indication that Truman was lukewarm to a summit. For example, Trevor Burridge states that there was an "absence of any American initiative," and Margaret Gowing observes that there was no particular urgency on the part of the United States to hold Anglo-American talks. A study of British sources does give this false impression. But American sources show the existence of the president's misplaced reply of October 5. Moreover, the ten days between September 25 and October 5 were well spent laying the foundation for a future summit. See Trevor Burridge, *Clement Attlee: A Political Biography* (London: Jonathan Cape, 1985), and Gowing, *Policy Making.*

31. "President to Prime Minister," October 17, 1945, Post-Presidential Files, box 3.

32. "The Future of the Quebec Agreement," November 3, 1945, PRO, FO 800/541.

33. "Relations with the United States," September 24, 1945, PRO, FO 800/537.

34. "Future Arrangement for Anglo-American Cooperation on Atomic Energy," n.d., PRO, FO 800/541.

35. "Nevile Butler to Sir A. Cardogan," November 5, 1945, PRO, FO 800/541.

36. House of Commons Debate, 5th series, vol. 414, col. 22, October 9, 1945, col. 1866, October 23, 1945; col. 2012, October 24, 1945.

37. "Attlee to Churchill," September 28, 1945, PRO, PREM 8/113.

38. "Churchill to Prime Minister," October 6, 1945, PRO, PREM 8/113.

39. "From Foreign Office to Washington," October 31, 1945, PRO, FO 800/568. In his diary, Dalton speculated that Mark Oliphant, the Australian scientist who had worked with the British team at Los Alamos, was responsible for the leak.

40. "Extract from American Radio Bulletin," November 1, 1945, PRO, FO 800/568.

41. "From Washington to Foreign Office," October 25, 1945, PRO, PREM 8/117.

42. "Negotiations with the British," April 17, 1946, General Records of the Department of State, box 56.

43. "JSM [Joint Staff Mission] Washington to Cabinet Office," November 5, 1945, PRO, FO 800/538.

44. "From Washington to Foreign Office," November 8, 1945, PRO, FO 800/538.

45. "Negotiations with the British," April 17, 1946, General Records of the Department of State, box 56; also in Gregg Herken, *The Winning Weapon: The Atomic Bomb in the Cold War, 1945–50* (New York: Alfred A. Knopf, 1980), 64.

46. Ibid.

47. Copies of the so-called Memorandum of Intention can be found in General Records of the Department of State, box 56; also in Harrison-Bundy Files, file 50.

48. "Diplomatic History of the Manhattan Project," n.d., Harrison-Bundy Files, file 111; also "Negotiations with the British," April 17, 1946, General Records of the Department of State, box 56.

49. "Diplomatic History of the Manhattan Project," Harrison-Bundy Files, file 111.

50. Ibid.

51. "Memorandum to the Secretary of War," November 23, 1945, Harrison-Bundy Files, file 5.

52. Kenneth Harris, *Attlee* (London: W. W. Norton, 1982), 282.

53. Ibid., 282–83.

54. "Cabinet Advisory Committee on Atomic Energy," December 4, 1945, PRO, FO 800/549.

55. "Diplomatic History of the Manhattan Project," Harrison-Bundy Files, file 111.

56. "JSM Washington to Cabinet Office," February 1, 1946, PRO, FO 800/527.

57. Roger Bullen and M. E. Pelly, eds., "Joint Staff Mission to Cabinet Office,"

February 17, 1946, no. 34, in *Documents on British Policy Overseas,* series 1, vol. 4 (London: Her Majesty's Stationery Office, 1987) (hereafter cited as Bullen and Pelly); also PRO FO 800/527.

58. David Lilienthal, *The Atomic Energy Years, 1945–1950,* vol. 2 of *The Journals of David E. Lilienthal* (New York: Harper and Row, 1964), 10.

59. Ibid., 25.

60. U.S. Department of State, "Groves to Secretary of State," February 13, 1946, in *Foreign Relations of the United States: General; the United Nations, 1946,* vol. 1, p. 1205 (hereafter cited as *FRUS, 1946,* vol. 1) (Washington, D.C.: Government Printing Office, 1972), 1204–7. Subsequent page-number references to this letter will appear in the text.

61. Bullen and Pelly, "Joint Staff Mission to Cabinet Office," February 17, 1946, no. 34; also FO 800/527; "Minutes of Meeting of Combined Policy Committee," February 15, 1946, *FRUS, 1946,* vol. 1, 1213–15.

62. Herken makes this argument in *Winning Weapon,* 145.

63. "Atomic Energy," December 8, 1945, PRO, FO 800/541.

64. Ibid.

65. "Cabinet Offices to Joint Staff Mission," December 20, 1945, PRO, FO 800/541.

66. "Joint Staff Mission to Cabinet Office," February 17, 1946, PRO, FO 800/527; also Bullen and Pelly, "Joint Staff Mission to Cabinet Office," February 17, 1946, no. 34.

67. Bullen and Pelly, "Joint Staff Mission to Cabinet Office," February 19, 1946, no. 36.

68. Bullen and Pelly, "Prime Minister to Lord Halifax," March 6, 1946, no. 45.

69. Ibid.; also "British Prime Minister to President Truman," April 15, 1946, *FRUS, 1946,* vol. 1, 1231–32.

70. Bullen and Pelly, "Lord Halifax to the Prime Minister," April 16, 1946, no. 70.

71. "Minutes of Meeting of CPC," April 15, 1946, *FRUS, 1946,* vol. 1, 1231.

72. Atlee sent a copy of the cable to Halifax: Bullen and Pelly, "Prime Minister to Lord Halifax," April 16, 1946, no. 72; also "Prime Minister to President Truman," April 16, 1946, *FRUS, 1946,* vol. 1, 1231–32.

73. Ibid., also "Prime Minister to President," April 16, 1946, Harrison-Bundy Files, file 103.

74. "President Truman to Mr. Attlee," April 20, 1946, President Secretary's Files, box 200; also *FRUS, 1946,* vol. 1, April 20, 1946, 1235–37.

75. Ibid.

76. "Attlee to Truman," June 6, 1946, President's Secretary's Files, box 200; also in *FRUS, 1946,* vol. 1, June 6, 1946, 1249–53, and Bullen and Pelly, June 6, 1946, no. 107.

77. Ibid.

78. Bullen and Pelly, "Lord Halifax to Prime Minister," April 19, 1946, no.

77; also "Memorandum of Conversation," April 18, 1946, General Records of the Department of State, box 57.

79. "Patterson to Burns," April 17, 1946, General Records of the Department of State, box 57.

80. Lilienthal, *Atomic Energy Years,* 23–24.

81. "Lord Halifax to Anderson," February 17, 1946, PRO, FO 800/527.

82. Lilienthal, *Atomic Energy Years,* 23–24.

83. Quoted in editorial footnote of Bullen and Pelly, "Lord Halifax to Prime Minister," April 19, 1946, no. 77.

84. Lilienthal, *Atomic Energy Years,* 10.

85. Bullen and Pelly, "Notes by Sir John Anderson," April 24, 1946, no. 80.

86. "Akers to Simon," August 22, 1945, War of 1939–1945: Correspondence and Papers (AB 1), Public Record Office, London (hereafter cited as PRO, AB 1), AB 1/606.

87. "Memorandum to Chancellor of the Exchequer," February 16, 1945, PRO, PREM 3/139/11A.

88. "Combined Policy Committee Meeting," October 13, 1945, Harrison-Bundy Files, file 38.

89. "Lord Halifax to John Anderson," February 17, 1946, PRO, FO 800/582.

90. "Revised Quebec Agreement," November 25, 1945, PRO, FO 800/541.

91. "Memorandum on Allocation by the British Members of the Combined Policy Committee," n.d., *FRUS, 1946,* vol. 1, 1225–26.

92. "Meeting of the Combined Policy Committee," April 15, 1946, *FRUS, 1946,* vol. 1, 1227–31.

93. He estimated that the Congo reserve of high-grade ore was 7,700 tons. U.S. needs for 1946 were 3,060 tons and 230 tons per month thereafter. British estimated requirements were 5,400 tons.

94. "Memorandum by the Commanding General, Manhattan Engineer District, to Acheson and Bush," April 29, 1946, *FRUS, 1946,* vol. 1, 1238–41.

95. "Rickett to Makins," April 25, 1946, PRO, FO 800/580.

96. "Lord Halifax to John Anderson," May 2, 1946, PRO, FO 800/580.

97. Ibid.

98. "Memorandum by John M. Hancock," May 1, 1946, *FRUS, 1946,* vol. 1, 1243.

99. "Acheson, Bush and Groves to the US Members of the CPC," May 7, 1946, *FRUS, 1946,* vol. 1, 1246.

100. R. Gordon Arneson, interview by Neil M. Johnson, June 21, 1989, transcript, Oral History Collection, Truman Library, Independence, Mo., interview 456.

CHAPTER FIVE

1. Gregg Herken, *The Winning Weapon: The Atomic Bomb in the Cold War, 1945–50* (New York: Alfred A. Knopf, 1980), 32.

2. "Popular Opinion on the Atomic Bomb as Revealed by Opinion Poll," Mid-October, 1945, Records of the Special Assistant to the Secretary of State for Atomic Energy Matters, Lot 57 D688, General Records of the Department of State, National Archives, Suitland Branch, Md. (hereafter cited as General Records of the Department of State), box 66.

3. James G. Hershberg, *James B. Conant: Harvard to Hiroshima and the Making of the Nuclear Age* (New York: Alfred A. Knopf, 1993), 215, 230; Martin J. Sherwin, *A World Destroyed: Hiroshima and the Origins of the Arms Race* (New York: Vintage Books, 1987), 90–128, 210–12.

4. "Atomic Control," November 10, 1945, McNaughton Reports, Papers of Frank McNaughton, September 15 to December 1945, Truman Library, Independence, Mo. (hereafter cited as Papers of Frank McNaughton), box 8.

5. "Newspaper Opinion on Control of Atomic Energy," August 8 to October 23, 1945, General Records of the Department of State, box 66.

6. Ibid.

7. "Memorandum for President," September 25, 1945, Harrison-Bundy Files, Records of the Office of the Chief of Engineers, Manhattan Engineer District (Record Group 77), National Archives, Washington, D.C. (hereafter cited as Harrison-Bundy Files), file 20.

8. "Cabinet Meeting," September 26, 1945, President's Secretary's Files, General File, Truman Library, Independence, Mo. (hereafter cited as President's Secretary's Files), box 112.

9. Quoted in James F. Byrnes, *Speaking Frankly* (New York: Harper and Brothers, 1947), 269.

10. John Morton Blum, ed., *The Price of Vision: The Diary of Henry A. Wallace, 1942–1946* (Boston: Houghton Mifflin, 1973), 486.

11. Walter Millis, ed., *The Forrestal Diaries* (New York: Viking Press, 1951), 95.

12. "Clinton P. Anderson to President Truman," September 25, 1945, President's Secretary's Files, box 199.

13. George F. Kennan, *Memoirs, 1925–1950* (Boston: Little, Brown, 1967), 296–97.

14. "Rickett to Makins," August 30, 1945, War of 1939–1945: Correspondence and Papers (AB 1), Public Record Office, London (hereafter cited as PRO, AB 1), PRO, AB 1/606.

15. Hugh Dalton, "Hugh Dalton Diary," October 17, 1945, Hugh Dalton Diaries and Papers, British Library of Economics, London School of Economics, London.

16. Ibid.

17. "Memorandum by the Prime Minister," November 5, 1945, Private Collections: Ministers and Officials: Various (FO 800), Public Record Office, London (hereafter cited as PRO, FO 800), PRO, FO 800/548.

18. "Memorandum by the Prime Minister," November 8, 1945, Ad-Hoc Committees: Gen and Miscellaneous Series (CAB 130), Public Record Office, London (hereafter cited as PRO, CAB 130), PRO, CAB 130/3.

19. "Memorandum by the Prime Minister," November 8, 1945, PRO, FO 800/548; also PRO, CAB 130/3.

20. "Chiefs of Staff Committee to Prime Minister," October 10, 1945, PRO, FO 800/547.

21. Quoted in Timothy J. Botti, *The Long Wait: The Forging of the Anglo-American Nuclear Alliance, 1945–1958* (New York: Greenwood Press, 1987), 7.

22. "Atomic Energy," Papers of Frank McNaughton, box 8.

23. Herken, *Winning Weapon,* 117–18; Hershberg, *James B. Conant,* 258–61.

24. Quoted in Herken, *Winning Weapon,* 119.

25. "Memorandum to Secretary of War and Secretary of Navy," November 28, 1945, Maj. Gen. Leslie R. Groves Files, Records of the Office of the Chief of Engineers, Manhattan Engineer District (Record Group 77), National Archives, Washington, D.C. (hereafter cited as Gen. Groves Files), file 13.

26. Harry S. Truman, *Years of Trial and Hope,* vol. 2 of *Memoirs* (New York: Doubleday, 1956), 3.

27. "Press Release," February 2, 1946, President's Secretary's Files, box 170.

28. Herken, *Winning Weapon,* 125–28.

29. Ibid., 129–30.

30. Margaret Gowing, *Policy Execution,* vol. 2 of *Independence and Deterrence: Britain and Atomic Energy, 1945–1952* (London: Macmillan, 1974), 141.

31. "Ambassador Harriman to the President of the United States," June 12, 1946, President's Secretary's Files, box 170.

32. "Minutes by Mr Beckett on McMahon Bill on Atomic Energy," May 10, 1946, PRO, FO 800/580.

33. "Atomic Energy," June 5, 1946, PRO, FO 800/580.

34. "Chadwick to Rickett," June 6, 1946, PRO, FO 800/580.

35. "Makins to Rickett," June 6, 1946, PRO, FO 800/580.

36. "Chadwick to Rickett," June 6, 1946, PRO, FO 800/580.

37. "Statement by Senator Brien McMahon," February 17, 1952, Records of the Joint Congressional Committee on Atomic Energy, 1946–77, General Correspondence, box 316A, Center for Legislative Archives, National Archives, Washington, D.C.

38. Dean Acheson, *Present at the Creation: My Years in the State Department* (New York: W. W. Norton, 1969), 167.

39. Ibid., 166.

40. Dalton, "Hugh Dalton Diary," October 5, 1945.

41. Kenneth Harris, *Attlee* (London: W. W. Norton, 1982), 273.

42. Ibid., 274.

43. Quoted in Harris, *Attlee,* 286.

44. David Lilienthal, *The Atomic Energy Years, 1945–50,* vol. 2 of *The Journals of David E. Lilienthal* (New York: Harper and Row, 1964), 218–19.

45. "A Report on the International Control of Atomic Energy," March 16, 1946, President's Secretary's Files, box 200.

46. Truman, *Years of Trial and Hope,* 7.

47. Lilienthal, *Atomic Energy Years,* 39.

48. Truman, *Years of Trial and Hope,* 10.

49. Bernard Baruch, *Baruch: The Public Years* (Holt, Rinehart and Winston, 1960), 364.

50. Ibid., 371.

51. Ibid., 372.

52. Byrnes, *Speaking Frankly,* 265.

53. Lilienthal, *Atomic Energy Years,* 58.

54. Ibid., 60.

55. Martin Walker, *The Cold War: A History* (New York: Henry Holt, 1993), 24.

56. Lincoln Gordon, interview by Richard D. McKenzie, transcript, July 17, 1975, Oral History Collection, Truman Library, Independence, Mo., interview 232.

57. Ibid.

58. Roger Bullen and M. E. Pelly, eds., "Mr Bevin to Sir A. Cadogan," March 22, 1946, no. 34, in *Documents on British Policy Overseas,* series 1, vol. 4 (hereafter cited as Bullen and Pelly) (London: Her Majesty's Stationery Office, 1987).

59. Bullen and Pelly, "Estimate of the Attitude of the Soviet Government towards the United Kingdom Proposals," March 20, 1946, no. 52.

60. Bullen and Pelly, "Mr Bevin to Sir A. Cadogan," July 5, 1946, no. 119.

61. House of Commons Debate, 5th series, vol. 426, col. 1359–87, August 2, 1946.

62. Lilienthal, *Atomic Energy Years,* 71.

63. "Note of a Meeting of Ministers," December 16, 1946, PRO, FO 800/585.

64. Baruch, *Baruch,* 379.

65. Report summarized in Millis, *Forrestal Diaries,* 217.

66. Summarized in Herken, *Winning Weapon,* 186–88.

67. Richard G. Hewlett and Francis Duncan, *Atomic Shield, 1947–1952,* vol. 2 of *A History of the United States Atomic Energy Commission* (Pittsburgh: Pennsylvania State University Press, 1969), 270–71.

68. U.S. Department of State, "Under Secretary of State to Ambassador-Belgium," June 27, 1947, *Foreign Relations of the United States: General; the United Nations, 1947,* vol. 1 (Washington, D.C.: Government Printing Office, 1973), 822–24.

CHAPTER SIX

1. Margaret Gowing, *Policy Making,* vol. 1 of *Independence and Deterrence: Britain and Atomic Energy, 1945–1952* (London: Macmillan, 1974), 19–20.

2. House of Commons Debate, 5th series, vol. 415, col. 38, October 29, 1945.

3. House of Commons Debate, 5th series, vol. 418, col. 682–84, January 29, 1946.

4. House of Commons Debate, 5th series, vol. 450, col. 2444–73, May 14, 1948.

5. Quoted in Kenneth Harris, *Attlee* (London: W. W. Norton, 1982), 288.

6. Roger Ruston, *A Say in the End of the World: Morals and British Nuclear Weapons Policy, 1941–1987* (Oxford, England: Oxford University Press, 1990), 91.

7. Quoted in Alan Bullock, *Ernest Bevin: Foreign Secretary, 1945–51* (London: W. W. Norton, 1983), 352.

8. "Meeting of Ministers," January 8, 1947, Ad-Hoc Committees: Gen and Miscellaneous Series (CAB 130), Public Record Office, London (hereafter cited as PRO, CAB 130), PRO, CAB 130/16.

9. Hugh Dalton, *High Tide and After: Memoirs, 1945–1960* (London: Frederick Miller, 1962), 187.

10. "Memorandum by the Minister of Supply," December 31, 1946, PRO, CAB 130/16.

11. "Cabinet Office to Joint Staff Mission Washington," September 30, 1946, Private Collections: Ministers and Officials: Various (FO 800), Public Record Office, London (hereafter cited as PRO, FO 800), PRO, FO 800/581.

12. "JSM Washington to Cabinet Offices," October 2, 1946, PRO, FO 800/581.

13. "Letter from Prime Minister to the President," December 17, 1946, Records of the Special Assistant to the Secretary of State for Atomic Energy Matters, Lot 57 D688, General Records of the Department of State, National Archives, Suitland Branch, Md. (hereafter cited as General Records of the Department of State), box 56.

14. "Letter from the President to the Prime Minister," n.d., General Records of Department of State, box 56.

15. "Records of Conversation," January 3, 1947, PRO, FO 800/597.

16. "JSM Washington to Minister of Defence," January 23, 1947, PRO, FO 800/597.

17. "Informal Memorandum," January 29, 1947, General Records of the Department of State, box 57; also "JSM Washington to Cabinet Office," January 31, 1947, PRO, FO 800/597.

18. [Untitled], February 11, 1947, General Records of the Department of State, box 56.

19. "Secretary of War to Secretary of State," n.d., General Records of the Department of State, box 56; also U.S. Department of State, "Secretary of War to Secretary of State," March 5, 1947, in *Foreign Relations of the United States: General; the United Nations, 1947,* vol. 1 (hereafter cited as *FRUS, 1947,* vol. 1) (Washington, D.C.: Government Printing Office, 1973), 798–99.

20. "Atomic Energy," February 25, 1946, PRO, FO 800/582.

21. "A Program of Negotiation with the British and Canadians," October 15, 1947, General Records of the Department of State, box 56.

22. "Chairman of the AEC to the Commissioner," April 23, 1947, *FRUS, 1947,* vol. 1, 804–806.

23. "JSM Washington to Cabinet Offices," February 4, 1947, PRO, FO 800/ 597.

24. "Atomic Energy," February 26, 1947, PRO, FO 800/613.

25. David Lilienthal, *The Atomic Energy Years, 1945–1950,* vol. 2 of *The Journals of David E. Lilienthal* (New York: Harper and Row, 1964), 175.

26. "Acheson to an Executive Session of the JCAE," May 12, 1947, *FRUS, 1947,* vol. 1, 806–11.

27. R. Gordon Arneson, interview by Neil M. Johnson, transcript, June 21, 1989, Oral History Collection, Truman Library, Independence, Mo., interview 456.

28. "B. B. Hickenlooper to Secretary of State," August 29, 1947, General Records of the Department of State, box 56; also *FRUS, 1947,* vol. 1, 833–34.

29. The existence of the wartime agreements would not become a part of the public record until June 1949, when Hickenlooper, then engaged in leveling charges of mismanagement against the AEC, requested as evidence the Quebec Agreement, the Hyde Park Aide-Memoire, and the Modus Vivendi.

30. "Gordon Munro to Rickett," May 31, 1947, Ministry of Supply Energy Division and London Office Files (AB 16), Public Record Office, London (hereafter cited as PRO, AB 16), PRO AB 16/388.

31. Ibid.

32. "Rickett to Makins," June 16, 1947, PRO, AB 16/388.

33. "JSM to Cabinet Office," August 25, 1947, PRO, AB 16/2022.

34. Ibid., August 27, 1947.

35. "A Program of Negotiation with the British and Canadians," October 15, 1947, General Records of the Department of State, box 56.

36. Gowing, *Policy Making,* 243.

37. "Gordon Munro to J. G. Stewart," September 5, 1947, PRO, AB 16/2017.

38. Ibid., October 2, 1947.

39. Richard G. Hewlett and Francis Duncan, *Atomic Shield, 1947–1952,* vol. 2 of *A History of the United States Atomic Energy Commission* (Pittsburgh: Pennsylvania State University Press, 1962), 147.

40. Ibid., 172–73.

41. Jonathan E. Helmreich, *Gathering Rare Ores: The Diplomacy of Uranium Acquisition, 1943–54* (Princeton, N.J.: Princeton University Press, 1986), 6.

42. Gowing, *Policy Making,* 367.

43. Hewlett and Duncan, *Atomic Shield,* 172–73.

44. Ibid., 54.

45. "Edmund Gullion to Acheson," March 26, 1947, *FRUS, 1947,* vol. 1, 802–4.

46. Ibid.

47. "Edmund Gullion to Acheson," June 9, 1947, *FRUS, 1947,* vol. 1, 835.

48. "Ambassador in Belgium to Under Secretary of State Lovett," September 2, 1947, *FRUS, 1947,* vol. 1, 835.

49. Lilienthal, *Atomic Energy Years,* 175.

50. Ibid., 182.

51. "Kennan to Lovett," October 24, 1947, *FRUS, 1947,* vol. 1, 843, footnote.

52. "Memorandum by Kennan and Gullion," October 24, 1947, *FRUS, 1947,* vol. 1, 844.

53. A notable absentee was General Groves, chairman of the Military Liaison Committee. General Eisenhower, Forrestal, and Lilienthal, concerned with his obstructionism, were looking for ways to replace him on the committee. He would shortly announce his retirement, effective on February 29, 1948.

54. "Minutes of Meeting of American Members of CPC," November 5, 1947, *FRUS, 1947,* vol. 1, 852–60.

55. "Meeting of American Members of CPC with Chairman of JCAE and Chairman of Senate Foreign Relations Committee," November 26, 1947, *FRUS, 1947,* vol. 1, 870–79.

56. Lilienthal, *Atomic Energy Years,* 260.

57. "Meeting of CPC with Chairmen of JCAE and Foreign Relations Committee," November 26, 1947, *FRUS, 1947,* vol. 1, 878.

58. Ibid, p. 879.

59. "A Program of Negotiation with the British and Canadians," October 15, 1947, General Records of the Department of State, box 56.

60. "Meeting of CPC with Chairmen of JCAE and Foreign Relations Committee," November 26, 1947, *FRUS, 1947,* vol. 1, 879.

61. "Munro to Cabinet Offices," November 15, 1947, PRO, AB 16/2017; also "US to Seek Rare Minerals from Europe," *Washington Post,* November 14, 1947, p. 1.

62. "Lord Inverchapel to Makins," December 2, 1947, PRO, AB 16/2022.

63. "Lord Inverchapel to Makins," December 3, 1947, PRO, AB 16/2022.

64. Ibid.

65. "Acting Secretary of State to Secretary of State," December 6, 1947, *FRUS, 1947,* vol. 1, 885–86.

66. "Ambassador in the United Kingdom to the Secretary of State," December 13, 1947, *FRUS, 1947,* vol. 1, 896–97.

67. Ibid, p. 897.

68. "Lord Inverchapel to Foreign Office," December 6, 1947, PRO, AB 16/2022.

69. "Makins to Lord Inverchapel," December 8, 1947, PRO, AB 16/2022.

70. "Makins to Munro," December 3, 1947, PRO, AB 16/2022.

71. "Cabinet Offices to Joint Staff Mission," December 10, 1947, PRO, AB 16/2022.

72. Quoted in Alan Bullock, *Ernest Bevin,* 482.

73. Guy Hartcup and T. E. Allibone, *Cockcroft and the Atom* (Bristol, England: Adam Hilger, 1984), 221.

74. "Meeting of the Combined Policy Committee," December 10, 1947, *FRUS, 1947,* vol. 1, 889–94.

75. Hartcup and Allibone, *Cockcroft and the Atom,* 22.

76. "Ambassador and Gordon Munro to Cabinet Offices," December 18, 1947, PRO, AB 16/2022.

77. "Teletype Conference between Counselor of the Department of State (Bohlen) and Kennan," December 17, 1947, *FRUS, 1947,* vol. 1, 905–6.

78. "Statement by the Undersecretary of State before the Joint Congressional Committee on Atomic Energy," January 21, 1948, *Foreign Relations of the United States: General; the United Nations, 1948,* vol. 1, pt. 2 (Washington, D.C.: Government Printing Office, 1973) (hereafter cited as *FRUS, 1948,* vol. 1, pt. 2), 690–91.

79. Hewlett and Duncan, *Atomic Shield,* 282.

80. "Minutes of Combined Policy Committee," January 7, 1948, *FRUS, 1948,* vol. 1, pt. 2, 679–86; also "Modus Vivendi," Papers of Harry S. Truman, Naval Aide File, Truman Library, Independence, Mo., box 13.

81. Hewlett and Duncan, *Atomic Shield,* 283.

82. Hartcup and Allibone, *Cockcroft and the Atom,* 221.

83. Gowing, *Policy Making,* 221.

84. Quoted in ibid.

85. Roger Bullen and M. E. Pelly, eds., "Conversation between Mr Bevin, Senator Vandenburg and M. J. Foster Dulles," January 26, 1946, no. 18, in *Documents on British Policy Overseas,* series 1, vol. 4 (London: Her Majesty's Stationery Office, 1987).

86. Walter Lafeber, *The American Age: United States Foreign Policy at Home and Abroad since 1750* (New York: W. W. Norton, 1989), 437.

87. Kenneth O. Morgan, *Labour in Power, 1945–1951* (Oxford, England: Clarendon Press, 1984), 273–74.

88. Ibid.

89. Ibid.

90. "Under Secretary of State before the JCAE," January 21, 1948, *FRUS, 1948,* vol. 1, pt. 2, 688–91.

91. "Memorandum of Conversation," January 7, 1948, *FRUS, 1948,* vol. 1, pt. 2, 677–78.

92. "Under Secretary of State before the JCAE," January 21, 1948, *FRUS, 1948,* vol. 1, pt. 2, 688–91.

93. "Makins to Foreign Office," January 13, 1948, PRO, AB 16/388.

94. "Memorandum of Conversation," September 16, 1948, *FRUS, 1948,* vol. 1, pt. 2, 755.

95. "F. N. Woodward to H. L. Verry," January 14, 1948, War of 1939–1945: Correspondence and Papers (AB 1), Public Record Office, London, PRO, AB 1/ 602.

CHAPTER SEVEN

1. "Review of Technical Collaboration in Atomic Energy Between US, UK and Canada," June 10, 1948, Ministry of Supply Energy Division and London

Office Files (AB 16), Public Record Office, London (hereafter cited as PRO, AB 16), PRO, AB 16/303.

2. The Americans were still coming up with one excuse after another to keep them out of Hanford, where most of the plutonium extraction work was done. The British resigned themselves to the likelihood that the Americans would not agree to let them in.

3. "Review of Technical Collaboration in Atomic Energy Between US, UK and Canada," June 10, 1948, PRO, AB 16/303.

4. "Meeting of the Chiefs of Staff," March 10, 1948, Chiefs of Staff Committee: Minutes of Meetings, January 1, 1947–December 28, 1950 (DEFE 4), Public Record Office, London (hereafter cited as PRO, DEFE 4), PRO, DEFE 4/11.

5. House of Commons Debate, 5th series, vol. 450, col. 2444–73, May 14, 1948.

6. "Meeting of the Chiefs of Staff," March 10, 1948, PRO, DEFE 4/11.

7. U.S. Department of State, "Memorandum of Conversation by Edmund Gullion," March 19, 1948, *Foreign Relations of the United States: General; the United Nations, 1948*, vol. 1, pt. 2 (Washington, D.C.: Government Printing Office, 1973) (hereafter cited as *FRUS, 1948*, vol. 1, pt. 2), 700.

8. "Note for the Ambassador," n.d., *FRUS, 1948*, vol. 1, pt. 2, 773–75.

9. David Lilienthal, *The Atomic Energy Years, 1945–1950*, vol. 2 of *The Journals of David E. Lilienthal* (New York: Harper and Row, 1964), 380.

10. Ibid., 380–81.

11. "Minutes of Meeting of American Members of CPC," July 6, 1948, *FRUS, 1948*, vol. 1, pt. 2, 719–23. Quote is on p. 721.

12. Ibid.

13. Lilienthal's comments in this paragraph are all from Lilienthal, *Atomic Energy Years*, 385.

14. "Memorandum by the Chairman of the Research and Development Board (Bush) to the Secretary of Defense (Forrestal)," August 12, 1948, Records of the Special Assistant to the Secretary of State for Atomic Energy Matters, Lot 57 D688, General Records of the Department of State, National Archives, Suitland Branch, Md. (hereafter cited as General Records of the Department of State), box 55; also in *FRUS, 1948*, vol. 1, pt. 2, 738–39.

15. Ibid.

16. "Arneson to Acting Secretary of State," September 27, 1948, *FRUS, 1948*, vol. 1, pt. 2, 767–70.

17. "Arneson to Lovett," November 2, 1948, *FRUS, 1948*, vol. 1, pt. 2, 781–84.

18. "Arneson to Acting Secretary of State," September 27, 1948, *FRUS, 1948*, vol. 1, pt. 2, 768.

19. "Memorandum of Meeting," August 12, 1948, *FRUS, 1948*, vol. 1, pt. 2, 734, 735.

20. "Makins to Cockcroft," June 9, 1948, PRO, AB 16/388.

21. "Memorandum of Meeting," August 12, 1948, *FRUS, 1948*, vol. 1, pt. 2, 734–37.

22. "Arneson to Lovett," November 2, 1948, *FRUS, 1948,* vol. 1, pt. 2, 781–84.

23. Ibid.

24. "Makins to Cockcroft," June 9, 1948, PRO, AB 16/388.

25. "Brigadier Price to Lord Portal," August 25, 1948, PRO, AB 16/269.

26. "Arneson to Acting Secretary of State," September 27, 1948, *FRUS, 1948,* vol. 1, pt. 2, 767–70.

27. "Copy of a Letter from General Hollis to the Chairman, Joint Services Mission," August 8, 1948, PRO, AB 16/269.

28. "Memorandum for the United States Secretary of National Defense from the Minister of Defence in the United Kingdom," September 1, 1948, PRO, AB 16/269.

29. "Memorandum by Donald Carpenter, Deputy to the Secretary of Defense," September 16, 1948, *FRUS, 1948,* vol. 1, pt. 2, 755–58.

30. "Arneson to Acting Secretary of State," September 27, 1948, *FRUS, 1948,* vol. 1, pt. 2, 768, footnote.

31. "Woodward to Cockcroft," September 20, 1948, PRO, AB 16/303.

32. "Note for the Ambassador's Talk with Mr. Lovett," n.d., *FRUS, 1948,* vol. 1, pt. 2, 774.

33. Ibid.

34. "Memo of Conversation," September 30, 1948, *FRUS, 1948,* vol. 1, pt. 2, 770–72.

35. Ibid.

36. "Cornwall-Jones to Michael Perrin," November 18, 1948, PRO, AB 16/269.

37. "Cornwall-Jones to Makins," November 23, 1948, PRO, AB 16/269.

38. "Cornwall-Jones to Michael Perrin," November 18, 1948, PRO, AB 16/269.

39. "Minutes of Meeting," October 4, 1948, PRO, DEFE 4/16.

40. Margaret Gowing, *Policy Making,* vol. 1 of *Independence and Deterrence: Britain and Atomic Energy, 1945–52* (London: Macmillan, 1974), 310.

41. "Memorandum of Conversation by Lovett," November 16, 1948, *FRUS, 1948,* vol. 1, pt. 2, 785–86.

42. Ibid.

CHAPTER EIGHT

1. U.S. Congress, Joint Congressional Committee on Atomic Energy, "Transcript of Executive Meeting," January to June 1949, Classified Transcripts of Meetings Held in Executive Session, Center for Legislative Archives, National Archives, Washington, D.C. (hereafter referred to as JCAE Classified Transcripts), doc. CDLXXII, box 002.

2. "Longair to Awberry," July 19, 1949, Ministry of Supply Energy Division and London Office Files (AB 16), Public Record Office, London (hereafter cited as PRO, AB 16), PRO, AB 16/276.

3. Richard G. Hewlett and Francis Duncan, *Atomic Shield, 1947–1952,* vol. 2 of *A History of the United States Atomic Energy Commission* (Pittsburgh: Pennsylvania State University Press, 1962), 295–96.

4. Dean Acheson, *Present at the Creation: My Years in the State Department* (New York: W. W. Norton, 1969), 164.

5. [Untitled], January 18, 1949, Records of the Special Assistant to the Secretary of State for Atomic Energy Matters, Lot 57 D688, General Records of the Department of State, National Archives, Suitland Branch, Md. (hereafter cited as General Records of the Department of State), box 57.

6. U.S. Department of State, "Memorandum by Mr. R. Gordon Arneson to Secretary of State and Under Secretary of State (Webb)," February 3, 1949, in *Foreign Relations of the United States: General; the United Nations, 1949,* vol. 1 (Washington, D.C.: Government Printing Office, 1976) (hereafter cited as *FRUS, 1949,* vol. 1), 421.

7. "Notes re Princeton Meeting," January 28, 1949, Papers of Harry S. Truman, Records of the National Security Council, Truman Library, Independence, Mo. (hereafter cited as Records of the National Security Council), box 18.

8. Ibid.

9. [Untitled], February 9, 1949, General Records of the Department of State, box 56.

10. "Memorandum Read to a Meeting of the Atomic Energy Commission on Saturday, 5 February, 1949, by Commissioner Lewis L. Strauss," February 5, 1949, General Records of the Department of State, box 56.

11. David Lilienthal, *The Atomic Energy Years, 1945–1950,* vol. 2 of *The Journals of David E. Lilienthal* (New York: Harper and Row, 1964), 464–67. First quote in this paragraph is from p. 464; the second is from p. 467.

12. "A Report to the President by Special Committee of National Security Council on Atomic Energy Policy with Respect to the UK and Canada," March 2, 1949, Records of the National Security Council, box 18; also in *FRUS, 1949,* vol. 1, March 2, 1949, 443–61.

13. Ibid.

14. Ibid.

15. "Roger Makins to Gordon Munro," February 8, 1949, PRO, AB 16/304.

16. "Atomic Energy Policy," March 9, 1949, General Records of the Department of State, box 56.

17. Ibid.

18. "Approach to the British on Atomic Energy," April 19, 1949, General Records of the Department of State, box 56.

19. "Memorandum of Conversation on General Norstad's Talks with Air Marshall Tedder," July 25, 1949, *FRUS, 1949,* vol. 1, 499–500.

20. "AEC Report," October 13, 1949, McNaughton Reports, Papers of Frank McNaughton, Truman Library, Independence, Mo., box 16.

21. "Note of Interview between Mr Carrol Wilson, Sir John Cockcroft and Dr Macfarlane," June 1, 1949, PRO, AB 16/304.

22. "Memorandum of Conversation on Tripartite Atomic Energy Negotiations," July 6, 1949, *FRUS, 1949,* vol. 1, 471–74.

23. The president resided at Blair House while the White House was under renovation.

24. The roll read as follows: the president and vice president, the secretary of state, the secretary of defense, Lilienthal, Eisenhower, Webster, Joseph Volpe of the AEC, and Gordon Arneson. Congress was represented by Speaker Sam Rayburn, Senators Tom Connally, Vandenberg, McMahon, Hickenlooper, and Millard Tydings and Representatives Carl Durham and Sterling Cole, both members of the JCAE.

25. "Record of Meeting at Blair House," July 14, 1949, *FRUS, 1949,* vol. 1, 476–82. This and the previous two quotes are from p. 479.

26. "Senator McMahon to Secretary of Defense," July 14, 1949, *FRUS, 1949,* vol. 1, 482–84.

27. Lilienthal, *Atomic Energy Years,* 545.

28. "Tripartite Negotiation Chronology," n.d., General Records of the Department of State, box 58.

29. "Memo of Telephone Conversation, Senator McMahon and Secretary Acheson," July 18, 1949, Papers of Dean Acheson, Truman Library, Independence, Mo. (hereafter cited as Papers of Dean Acheson), box 64.

30. "Arneson to Secretary of State," September 13, 1949, *FRUS, 1949,* vol. 1, 526–28.

31. The following account of this meeting is based on these sources: "Transcript of Executive Meeting," July 13 to October 10, 1949, JCAE Classified Transcripts, doc. 117, box 003; also *FRUS, 1949,* vol. 1, July 20, 1949, 490–98.

32. Lilienthal, *Atomic Energy Years,* 550.

33. Hewlett and Duncan, *Atomic Shield,* 304.

34. "Transcript of Executive Meeting," July 13 to October 10, 1949, JCAE Classified Transcripts, doc. 117, box 003; also in *FRUS, 1949,* vol. 1, July 20, 1949, 490–98.

35. "Tripartite Negotiation Chronology," June 23 to September 1, 1949, General Records of the Department of State, box 58.

36. Acheson, *Present at the Creation,* 319.

37. Ibid.

38. Lilienthal, *Atomic Years,* 543.

39. "Ernest A. Gross to Gordon Arneson," August 11, 1949, General Records of the Department of State, box 58.

40. "Memorandum for the Secretary," August 18, 1949, General Records of the Department of State, box 58.

41. "Tripartite Negotiation Chronology," June 23 to September 1, 1949, General Records of the Department of State, box 58.

42. Lilienthal, *Atomic Energy Years,* 555.

43. "Memorandum by Under Secretary of State (Webb)," August 19, 1949, *FRUS, 1949,* vol. 1, 519.

44. "Memorandum for Mr. Clark Clifford," July 27, 1949, Papers of Clark Clifford, Atomic Energy, Truman Library, Independence, Mo., box 1; also *FRUS, 1949,* vol. 1, July 27, 1949, 505–6.

45. "Transcript of Executive Meeting," July 13 to October 10, 1949, JCAE Classified Transcripts, doc. 100, box 003; also *FRUS, 1949,* vol. 1, July 27, 1949, 503–5.

46. "Memo of Conversation with the President," August 18, 1949, Papers of Dean Acheson, box 64; also "Memorandum of Conversations with the President," August 18, 1949, General Records of the Office of the Executive Secretariat, lot RG59, General Records of the Department of State, National Archives, Washington, D.C. (hereafter cited as General Records of the Office of the Executive Secretariat), box 1; also *FRUS, 1949,* vol. 1, August 18, 1949, 516.

47. "Tripartite Negotiation Chronology," June 23 to September 1, 1949, General Records of the Department of State, box 58.

48. Ibid.; also *FRUS, 1949,* vol. 1, August 12, 1949, 513–14.

49. Lilienthal, *Atomic Energy Years,* 565.

50. Representing the British were the following five knights: Sir Oliver Franks and Sir Derick Hoyer Millar, permanent members of the CPC; Sir John Cockcroft and the recently knighted Sir Roger Makins, both of whom had flown in from England for the purpose; General Sir William Morgan, army representative on the Joint Staff Mission; and two embassy attaches, A. K. Longair and Dr. W. A. Macfarlane.

51. "Minutes of Meeting of CPC," September 20, 1949, *FRUS, 1949,* vol. 1, 531–35.

52. Margaret Gowing, *Policy Making,* vol. 1 of *Independence and Deterrence: Britain and Atomic Energy, 1945–1952* (London: Macmillan, 1974), 284–88.

53. "Conclusions Concerning Tripartite Talks," September 27, 1949, General Records of the Department of State, box 58.

54. "Memorandum of Conversation of Meeting with the President," October 1, 1949, *FRUS, 1949,* vol. 1, 543.

55. "Memorandum of Conversation with the President," October 3, 1949, General Records of the Office of the Executive Secretariat, box 1; also *FRUS, 1949,* vol. 1, October 3, 1949, 552–53.

56. "Public Opinion News Service," September 8, 1949, General Records of the Department of State, box 58.

57. Harry S. Truman, *Years of Trial and Hope,* vol. 2 of *Memoirs* (New York: Doubleday, 1956), 306.

58. "Status of the USSR Atomic Energy Project," July 1, 1949, President's Secretary's Files, National Security Council, Truman Library, Independence, Mo., box 258.

59. Walter Millis, ed., *The Forrestal Diaries* (New York: Viking Press, 1951), 495–96.

60. "Memorandum of Conversation with Professor Niels Bohr," January 27, 1948, *FRUS, 1949,* vol. 1, pt. 2, 508.

61. Quoted in Kenneth Harris, *Attlee* (London: W. W. Norton, 1982), 290.

62. "Arneson to Webb," October 6, 1949, *FRUS, 1949,* vol. 1, 558.

63. "Transcript of Executive Meeting," October 11, 1949, to February 3, 1950, JCAE Classified Transcripts, doc. 65, box 004.

64. "Meeting with the President," October 13, 1949, Papers of Dean Acheson, box 64.

65. "Transcript of Executive Meeting," July 13 to October 10, 1949, JCAE Classified Transcripts, doc. 88, box 003.

66. "Discussion with Senator Vandenberg Concerning Atomic Energy Developments," September 23, 1949, General Records of the Department of State, box 08.

67. "The Acheson-Makins Talks," *Washington Post,* October 1, 1949, p. 1.

68. "Copy of a Minute from Sir Henry Tizard to the Minister," September 26, 1949, Prime Minister's Office: Correspondence and Papers, 1945–51 (PREM 8), Public Record Office, London (hereafter cited as PRO, PREM 8), PRO, PREM 8/1101.

69. "Prime Minister," October 17, 1949, PRO, PREM 8/1098.

70. Gowing, *Policy Making,* 293.

71. "Arneson to Senator McMahon," December 16, 1949, General Records of the Department of State, box 56.

72. James Reston, "Britain, Canada to Keep Up Supply of Uranium to US," *New York Times,* October 30, 1949, p. 1.

73. "Tripartite Talks," December 1, 1949, PRO, AB 16/304.

74. "For Makins from Oliver Franks," December 18, 1949, PRO, AB 16/304.

75. "Arneson to Senator McMahon," December 16, 1949, General Records of the Department of State, box 56.

76. "Remarks by Carrol Wilson in the Steering Group on Dec. 2nd," December 12, 1949, PRO, AB 16/268.

77. "BJSM to Cabinet Offices," December 12, 1949, PRO, AB 16/304.

78. "BJSM to Cabinet Office," December 3, 1949, PRO, AB 16/304; also "Atomic Energy Commission Tripartite Talks: Memorandum by Sir John Cockcroft," December 6, 1949, General Records of the Department of State, box 56.

79. "To Cabinet Office from BJSM," December 3, 1949, PRO, AB 16/304.

80. "Atomic Partnership Speculation Revived," *Washington Star,* December 7, 1949, p. 1.

81. Gowing, *Policy Making,* 297.

82. "Cabinet Offices to JSM," December, 22, 1949, PRO, AB 16/304.

83. "UK Draft Proposal for Future Tripartite Cooperation," December 29, 1949, PRO, AB 16/304.

84. Ibid.

85. "JSM to Cabinet Office," January 10, 1950, PRO, AB 16/858.

86. Gowing, *Policy Making,* 298.

87. "Memorandum by Messrs. Adrian S. Fisher and R. Gordon Arneson to

the Secretary of State," January 18, 1950, General Records of the Department of State, box 56; also U.S. Department of State, *Foreign Relations of the United States: General; the United Nations, 1950,* vol. 1 (Washington, D.C.: Government Printing Office, 1977) (hereafter cited as *FRUS, 1950,* vol. 1), 499–503.

88. "BJSM to Cabinet Office," December 12, 1949, PRO, AB 16/304.

89. "Fisher and Arneson to the Secretary of State," January 18, 1950, General Records of the Department of State, box 56; also *FRUS, 1950,* vol. 1, 499–503.

90. "JSM to Cabinet Office," January 10, 1950, PRO, AB 16/858.

CHAPTER NINE

1. The program would not be fully independent because the proposal called for the United States to supply the United Kingdom with enriched uranium 235 in exchange for British plutonium.

2. "Memorandum for the Secretary of State," January 18, 1950, Records of the Special Assistant to the Secretary of State for Atomic Energy Matters, Lot 57 D688, General Records of the Department of State, National Archives, Suitland Branch, Md. (hereafter cited as General Records of the Department of State), box 56.

3. "Tripartite Talks on Security Standards at Washington," June 19, 1950, General Records of the Department of State, box 58; see also John Costello, *Mask of Treachery* (London: Collins Sons, 1988), 521–28.

4. Costello, *Mask of Treachery,* 522.

5. Quoted in Kenneth Harris, *Attlee* (London: W. W. Norton, 1982), 200.

6. U.S. Department of State, "The Secretary of Defense (Johnson) to President," February 24, 1950, in *Foreign Relations of the United States: National Security Affairs; Foreign Economic Policy, 1950,* vol. 1 (Washington, D.C.: Government Printing Office, 1977) (hereafter cited as *FRUS, 1950,* vol. 1), 538–39.

7. "Suggested Policy Guidance in South African Discussions," March 13, 1950, *FRUS, 1950,* vol. 1, 542–43.

8. "Secretary of State to Secretary of Defense," April 3, 1950, *FRUS, 1950,* vol. 1, 546–47.

9. "Minutes of Meeting of American Members of CPC," April 25, 1950, *FRUS, 1950,* vol. 1, 547–50.

10. "Memorandum to the Combined Policy Committee," April 17, 1950, Department of State, box 58.

11. "BJSM in Washington to Cabinet Office," March 6, 1950, Ministry of Supply Energy Division and London Office Files (AB 16), Public Record Office, London (hereafter cited as PRO, AB 16), PRO, AB 16/858.

12. "Tripartite Talks on Security Standards at Washington," October 7, 1950, General Records of the Department of State, box 58.

13. Margaret Gowing, *Policy Execution,* vol. 2 of *Independence and Deterrence: Britain and Atomic Energy, 1945–1952* (London: Macmillan, 1974), 305.

14. After the Conservatives regained office, they agreed to introduce a questionnaire to be completed by all current staff and new entrants. Some 14,000 staff, officials, and scientists were involved. In addition, new applicants were required to provide the names of two referees. This new procedure, however, was implemented too slowly for American satisfaction.

15. "A. K. Longair to Sir John Cockcroft," July 17, 1950, PRO, AB 16/858.

16. "A. K. Longair to Sir John Cockcroft," August 9, 1950, PRO, AB 16/858.

17. Ibid.

18. "Memorandum by Lucius D. Battle, Special Assistant to Secretary of State," February 13, 1950, *FRUS, 1950,* vol. 1, 527–28.

19. "W. A. Macfarlane to D. E. H. Peirson," March 13, 1950, PRO, AB 16/858.

20. One of several processes under study in the United States, Canada, and the United Kingdom for separating various substances from irradiated uranium and thorium during the operation of a reactor.

21. "Frederick T. Hobbs to W. A. Macfarlane," May 23, 1950, PRO, AB 16/858.

22. "W. A. Macfarlane to D. E. H. Peirson," March 13, 1950, PRO, AB 16/858.

23. "Memorandum of Conversation Held at Mr. Bevin's Apartment," May 16, 1950, General Records of the Department of State, box 56.

24. "Cabinet Office to BJSM Washington," May 25, 1950, PRO, AB 16/2019; see also "Oliver Franks to Dean Acheson," June 22, 1950, General Records of the Department of State, box 58.

25. "Status of Tripartite Atomic Energy Relationship," June 28, 1950, General Records of the Department of State, box 56.

26. Richard G. Hewlett and Francis Duncan, *Atomic Shield, 1947–1952,* vol. 2 of *A History of the United States Atomic Energy Commission* (Pittsburgh: Pennsylvania State University Press, 1962), 410.

27. Ibid., 521.

28. Ibid., 524–25.

29. "Minutes of Meeting of American Members of CPC," September 7, 1950, *FRUS, 1950,* vol. 1, 572–75.

30. "Memorandum of Conversation," September 26, 1950, General Records of the Department of State, box 56.

31. Costello, *Mask of Treachery,* 528.

32. Ibid.

33. "GOP Cool to Easing Atomic Data Curb," October 1, 1951, PRO, AB 16/858.

34. U.S. Department of State, "Memorandum by the Secretary of Defense (Lovett), to the Chairman of the Combined Policy Committee (Acheson)," October 12, 1951, in *Foreign Relations of the United States: National Security Affairs; Foreign Economic Policy, 1951,* vol. 1 (Washington, D.C.: Government Printing Office, 1979) (hereafter cited as *FRUS, 1951,* vol. 1), 777.

35. "Statement Prepared by the Atomic Energy Commission," May 18, 1951, *FRUS, 1951,* vol. 1, 718–29.

36. Gowing, *Policy Execution,* 301.

37. "Press Release," November 30, 1950, Papers of Eben A. Ayers, box 4, Subject File, Truman Library, Independence, Mo.

38. Quoted in Harry S. Truman, *Years of Trial and Hope,* vol. 2 of *Memoirs* (New York: Doubleday, 1956), 396.

39. [Untitled], November 30, 1950, Lord Attlee's Papers, Bodleian Library, Oxford, MS114, fol. 154.

40. [Untitled], November 30, 1950, Lord Attlee's Papers, Bodleian Library, Oxford, MS114, fol. 159.

41. Roger Bullen and M. E. Pelly, eds., "Mr Bevin to Sir Franks," November 30, 1950, no. 82, in *Documents on British Policy Overseas,* series 2, vol. 4 (London: Her Majesty's Stationery Office, 1987) (hereafter cited as Bullen and Pelly).

42. "Record of a Meeting of the Prime Minister and Foreign Secretary with the French Prime Minister and Minister of Foreign Affairs," December 2, 1950, Private Collections: Ministers and Officials: Various (FO 800), Public Record Office, London (hereafter cited as PRO, FO 800), PRO, FO 800/465; see also Bullen and Pelly, no. 85.

43. "Foreign Office to Washington," August 14, 1950, Cabinet Office: Registered Files (CAB 21), Public Record Office, London (hereafter cited as PRO, CAB 21), PRO, CAB 21/1783.

44. "From Washington to Foreign Office," August 16, 1950, PRO, CAB 21/1783.

45. "Visit of Prime Minister Attlee to the United States," November 30, 1950, Papers of Dean Acheson, Truman Library, Independence, Mo. (hereafter cited as Papers of Dean Acheson), box 65.

46. Truman, *Memoirs,* 402. An account of the summit can be found in *Foreign Relations of the United States: Western Europe, 1950,* vol. 3 (Washington, D.C.: Government Printing Office, 1977) (hereafter cited as *FRUS, 1950,* vol. 3), 1706–87.

47. Dean Acheson, *Present at the Creation: My Years in the State Department* (New York: W. W. Norton, 1969), 484.

48. R. Gordon Arneson, interview by Neil M. Johnson, transcript, June 21, 1989, Oral History Collection, Truman Library, Independence, Mo., interview 456, 70–72.

49. "Truman-Attlee Joint Communiqué," December 8, 1950, President's Secretary's Files, Subject File, Truman Library, Independence, Mo. (hereafter cited as President's Secretary's Files), box 164.

50. Bullen and Pelly, "Minute from Sir Roger Makins to Mr Bevin," January 19, 1951, no. 111.

51. Ibid.

52. "Britons Told Wartime Controls Will Return," *Washington Post,* January 31, 1951, p. 1.

53. "Memorandum of Conversation on Personal Message from Mr. Bevin to Mr. Acheson," January 14, 1951, *FRUS, 1951,* vol. 1, 802–5.

54. "Memorandum of Conversation on Message from Mr. Bevin to the Secretary," January 15, 1951, *FRUS, 1951,* vol. 1, 805–6.

55. R. H. Shackford, "Britain Doesn't Like Role as Junior Atom Partner," *Washington Daily News,* February 1, 1951, p. 1.

56. House of Commons Debate, 5th series, vol. 483, col. 715–16, January 30, 1951.

57. "Acheson to Deputy to the Secretary of Defense (Lebaron)," February 9, 1951, *FRUS, 1951,* vol. 1, 690–91.

58. "Mr. Winston Churchill to President Truman," February 12, 1951, President's Secretary's Files, box 200; see also *FRUS, 1951,* February 12, 1951, 693–94.

59. "Prime Minister to Churchill," February 14, 1951, PRO, FO 800/438.

60. "President Truman to Mr. Winston Churchill," March 24, 1951, President's Secretary's Files, box 200; see also *FRUS, 1951,* March 24, 1951, 703–4.

61. House of Lords Debate, 5th. ser., vol. 172, col. 670–79, July 5, 1951.

62. Quoted in Kenneth O. Morgan, *Labour in Power, 1945–51* (Oxford, England: Clarendon Press, 1985), 465.

63. Ibid.

64. "Memorandum of Meeting between British Ambassador and Secretary Acheson," April 2, 1951, Papers of Dean Acheson, box 66.

65. Morgan, *Labour in Power,* 469.

66. "Informal Statement by Department of Defense," n.d. [probably September 14, 1951], *FRUS, 1951,* vol. 1, 771; see also "Secretary of Defense to Chairman CPC (Acheson)," October 12, 1951, *FRUS, 1951,* vol. 1, 777.

67. "Informal Statement by Department of Defense," n.d. [probably September 14, 1951], *FRUS, 1951,* vol. 1, 771; see also "Secretary of Defense to Chairman CPC (Acheson)," October 12, 1951, *FRUS, 1951,* vol. 1, 777.

68. Hewlett and Duncan, *Atomic Shield,* 549–50.

69. Ibid., 551–52.

70. "Memorandum for the File of Arneson," August 8, 1951, *FRUS, 1951,* vol. 1, 750–52.

71. "Dean to Acting Secretary of Defense (Foster)," November 27, 1951, *FRUS, 1951,* vol. 1, 787.

72. "F. W. Marten to Sir Roger Makins," September 18, 1951, PRO, AB 16/858.

73. Ibid.

74. "Memorandum on Discussion with the British," August 6, 1951, *FRUS, 1951,* vol. 1, 875–80.

75. "Memorandum of Conversation," September 11, 1951, Papers of Dean Acheson, box 66; see also *FRUS, 1951,* vol. 1, 880–83.

76. "United States Draft Statement," September 12, 1951, *FRUS, 1951,* vol. 1, 894.

77. "Memorandum of Discussion with British," August 6, 1951, *FRUS, 1951,* vol. 1, 875–80.

78. "Ministry of Defence to BJSM Washington," July 27, 1951, PRO, AB 16/269.

79. "N. Pritchard to Sir Alexander Clutterbuck," August 16, 1951, PRO, AB 16/269.

80. "Meeting of American Members of CPC," August 24, 1951, *FRUS, 1951,* vol. 1, 755–63.

81. "BJSM to Cabinet Office," September 7, 1951, PRO, AB 16/269.

82. "Cabinet Offices to BJSM," October 18, 1951, PRO, AB 16/867.

83. "Atomic Cooperation with the United States," September 21, 1951, PRO, AB 16/858.

84. "BJSM to Cabinet Office," December 4, 1951, PRO, AB 16/858.

85. "Official Committee on Atomic Energy," October 17, 1951, PRO, AB 16/858.

86. Ibid.

87. Ibid.

88. Ibid.

CHAPTER TEN

1. Margaret Gowing, *Policy Making,* vol. 1 of *Independence and Deterrence: Britain and Atomic Energy, 1945–1952* (London: Macmillan, 1974), 406.

2. Quoted in Gowing, *Policy Making,* 408.

3. U.S. Department of State, "The Ambassador in the United Kingdom (Gifford) to the Department of State," December 28, 1951, in *Foreign Relations of the United States: Western Europe and Canada, 1952–1954,* vol. 6, pt. 1 (Washington, D.C.: Government Printing Office, 1986) (hereafter cited as *FRUS, 1952–1954,* vol. 6, pt. 1), 720–23.

4. U.S. Department of State, "Substance of Discussion of State–Joint Chiefs of Staff Meeting," November 21, 1951, *Foreign Relations of the United States: National Security Affairs; Foreign Economic Policy; 1951,* vol. 1 (Washington, D.C.: Government Printing Office, 1979) (hereafter cited as *FRUS, 1951,* vol. 1), 898–99.

5. House of Commons Debate, 5th series, vol. 494, col. 376, November 21, 1951.

6. "Memorandum by McMahon on Churchill's Visit," December 5, 1951, *FRUS, 1951,* vol. 6, pt. 1, 695–96.

7. "The British Ambassador (Franks) to the Secretary of State," December 26, 1951, Records of the Special Assistant to the Secretary of State for Atomic Energy Matters, Lot 57 D688, General Records of the Department of State, National Archives, Suitland Branch, Md. (hereafter cited as General Records of the Department of State), box 59; also in *FRUS, 1951,* vol. 1, December 26, 1951, 798–99.

8. "UK Non-Acceptance of US Counter-Proposal on the Use of Nevada Test

Range for UK Atomic Weapon," January 3, 1952, General Records of the Department of State, box 59.

9. "Acting Secretary of State to UK Embassy," November 21, 1951, *FRUS, 1951*, vol. 6, pt. 1, 698–99.

10. "Memorandum of Conversation by Under Secretary of State," December 10, 1951, *FRUS, 1951*, vol. 6, pt. 1, 702–3.

11. Gowing, *Policy Making*, 411.

12. "Minutes of Second Formal Meeting of President Truman and Prime Minister Churchill," January 7, 1951, *FRUS, 1951*, vol. 6, pt. 1, 763–66.

13. "Record of a Meeting between Lord Cherwell and the Atomic Energy Commission," January 10, 1952, Ministry of Supply Energy Division and London Office Files (AB 16), Public Record Office, London (hereafter cited as PRO, AB 16), PRO, AB 16/860.

14. "Arneson to Secretary of State," January 15, 1952, in *Foreign Relations of the United States: Western Europe and Canada, 1952–1954*, vol. 2, pt. 2 (Washington, D.C.: Government Printing Office, 1986), 846–48. Quote is from p. 847.

15. "Atomic Energy," January 30, 1952, PRO, AB 16/1318.

16. Ibid.

17. Gowing, *Policy Making*, 414–15.

18. Ibid., 413–14.

19. "A. K. Longair to R. H. Fiskenden," February 6, 1952, PRO, AB 16/859.

20. Ibid.

21. "A. K. Longair to Dr W. A. Macfarlane," February 19, 1952, PRO, AB 15/859.

22. Ibid.

23. "Cabinet Office to BJSM," February 27, 1952, PRO, AB 16/867.

24. "The Tripartite Technical Cooperation Programme, Memorandum by the Ministry of Supply," November 28, 1952, PRO, AB 16/1318.

25. "Note of a Meeting at the Atomic Energy Commission," March 25, 1952, PRO, AB 16/861.

26. "Notes on Talks to AEC," May 6, 1952, PRO, AB 16/859.

27. Quoted in Gowing, *Policy Making*, 441.

28. "Gordon Arneson to James Penfield, American Embassy," October 10, 1952, General Records of the Department of State, box 58.

29. "Prime Minister's Statement on British Atomic Bomb Test," October 23, 1952, General Records of the Department of State, box 59.

30. Ibid.

31. Charles F. Barrett, "Move to Share US Atomic Data with Britain Faces Rough Fight," *Washington Star*, January 2, 1953, p. 1.

32. "Memorandum of Conversation by Sir Roger Makins," November 6, 1952, PRO, AB 16/1318; see also "Tripartite Technical Cooperation Programme," November 28, 1952, PRO, AB 16/1318.

EPILOGUE

1. Margaret Gowing, *Policy Making,* vol. 1 of *Independence and Deterrence: Britain and Atomic Energy* (London: Macmillan, 1974), 448–49.

2. Jan Melissen, *The Struggle for Nuclear Partnership: Britain, the United States and the Making of an Ambiguous Alliance, 1952–1959* (Groningen, the Netherlands: Styx Publications, 1993), 43.

3. Ibid.

4. "Memorandum to Dr Conant," September 23, 1944, Vannevar Bush–James B. Conant Files, fol. 19A, Office of Scientific Research and Development (Record Group 227), National Archives, Washington, D.C.

Bibliography

PRIMARY SOURCES

Unpublished Documents

Ad-Hoc Committees: GEN and Miscellaneous Series (CAB 130). Public Record Office, London.

Cabinet Memoranda from 1945 (CAB 129). Public Record Office, London.

Cabinet Minutes from 1945 (CAB 128). Public Record Office, London.

Cabinet Office: Registered Files (CAB 21). Public Record Office, London.

Chiefs of Staff Committee: Minutes of Meetings. January 1, 1947–December 28, 1950 (DEFE 4). Public Record Office, London.

Chiefs of Staff Committee: Minutes of Meetings. Pre-1947 (CAB 79). Public Record Office, London.

Harrison-Bundy Files. Records of the Office of the Chief of Engineers. Manhattan Engineer District (Record Group 77). National Archives, Washington, D.C.

Joint Congressional Committee on Atomic Energy. Classified Transcripts of Meetings Held in Executive Session. Center for Legislative Archives, National Archives, Washington, D.C.

Lyman J. Briggs Alphabetical Files. Office of Scientific Research and Development (Record Group 227). National Archives, Washington, D.C.

Maj. Gen. Leslie R. Groves Files. Records of the Office of the Chief of Engineers, Manhattan Engineer District (Record Group 77). National Archives, Washington, D.C.

Ministry of Supply Energy Division and London Office Files (AB 16). Public Record Office, London.

Operations Papers (PREM 3). Public Record Office, London.

Prime Minister's Office: Correspondence and Papers, 1945–51 (PREM 8). Public Record Office, London.

Private Collections: Ministers and Officials: Various (FO 800). Public Record Office, London.

Records of the Special Assistant to the Secretary of State for Atomic Energy Matters. Lot 57 D688. General Records of the Department of State. National Archives, Suitland, Md.

Vannevar Bush–James B. Conant Files. Office of Scientific Research and Development (Record Group 227). National Archives, Washington, D.C.

War of 1939–1945: Correspondence and Papers (AB 1). Public Record Office, London.

Published Documents

Bullen, Roger, and M. E. Pelly, eds. 1987. *Documents on British Policy Overseas.* London: Her Majesty's Stationery Office.

House of Commons Debates, 5th ser., vols. 415, 416, 418, 426, 450, 483, 494.

House of Lords Debates, vol. 172.

Kimball, Warren F., ed. 1984. *Alliance Forged: November 1942–February 1944.* Vol. 2 of *Churchill and Roosevelt: The Complete Correspondence.* Princeton, N.J.: Princeton University Press.

Stoff, Michael B., Jonathan F. Fanton, and R. Hal Williams, eds. 1991. *The Manhattan Project: A Documentary Introduction to the Atomic Age.* New York: McGraw-Hill.

U.S. Department of State. 1967. *Foreign Relations of the United States: General; Economic and Social Matters, 1944.* Vol. 2. Washington, D.C.: Government Printing Office.

———. 1972. *Foreign Relations of the United States: General; the United Nations, 1946.* Vol. 1. Washington, D.C.: Government Printing Office.

———. 1976. *Foreign Relations of the United States: General; the United Nations, 1949.* Vol. 1. Washington, D.C.: Government Printing Office.

———. 1977. *Foreign Relations of the United States: General; the United Nations, 1950.* Vol. 1. Washington, D.C.: Government Printing Office.

———. 1973. *Foreign Relations of the United States: General; the United Nations, 1947.* Vol. 1, pt. 2. Washington, D.C.: Government Printing Office.

———. 1976. *Foreign Relations of the United States: General; the United Nations, 1948.* Vol. 1, pt. 2. Washington, D.C.: Government Printing Office.

———. 1977. *Foreign Relations of the United States: Western Europe, 1950.* Vol. 3. Washington, D.C.: Government Printing Office.

———. 1979. *Foreign Relations of the United States: National Security Affairs; Foreign Economic Policy, 1951.* Vol. 1. Washington, D.C.: Government Printing Office.

———. 1984. *Foreign Relations of the United States: Western Europe and Canada, 1952–1954.* Vol. 2, pt. 2. Washington, D.C.: Government Printing Office.

———. 1986. *Foreign Relations of the United States: Western Europe and Canada, 1952–1954.* Vol. 6, pt. 1. Washington, D.C.: Government Printing Office.

Unpublished Private Papers

Churchill Papers. Churchill College, Cambridge, England.

Earl of Halifax Papers. Churchill College, Cambridge, England.

Lord Attlee's Papers. Bodleian Library, Oxford, England, and Churchill College, Cambridge, England.

Papers of Dean Acheson. Truman Library, Independence, Mo.

Papers of Eben A. Ayers. Truman Library, Independence, Mo.

Papers of Frank McNaughton. Truman Library, Independence, Mo.

Papers of George M. Elsey. Truman Library, Independence, Mo.

Papers of Harry S. Truman. Truman Library, Independence, Mo.

Papers of Mathew J. Connelly. Truman Library, Independence, Mo.

President's Secretary's Files. Truman Library, Independence, Mo.

Diaries and Memoirs

Acheson, Dean. 1969. *Present at the Creation: My Years in the State Department.* New York: W. W. Norton.

Anders, Roger M., ed. 1987. *Forging the Atomic Shield: Excerpts from the Office Diary of Gordon E. Dean.* Chapel Hill: University of North Carolina Press.

Baruch, Bernard. 1960. *Baruch: The Public Years.* New York: Holt, Rinehart and Winston.

Blum, John Morton, ed. 1973. *The Price of Vision: The Diary of Henry A. Wallace, 1942–1946.* Boston: Houghton Mifflin.

Bohlen, Charles. 1973. *Witness to History.* New York: W. W. Norton.

Byrnes, James F. 1947. *Speaking Frankly.* New York: Harper and Brothers.

Churchill, Winston. 1950. *The Hinge of Fate.* Vol. 4 of *The Second World War.* Boston: Houghton Mifflin.

Dalton, Hugh. "Hugh Dalton Diary." Hugh Dalton Diaries and Papers. British Library of Economics, London School of Economics.

Dalton, Hugh. 1962. *High Tide and After: Memoirs, 1945–1960.* London: Frederick Miller.

Ferrell, Robert H., ed. 1991. *Truman in the White House: The Diary of Eben A. Ayers.* Columbia, Mo.: University of Missouri Press.

Goldschmidt, Bertrand. 1990. *Atomic Rivals.* Translated by Georges M. Temmer. New Brunswick, N.J.: Rutgers University Press.

Groves, Leslie. 1962. *Now It Can Be Told: The Story of the Manhattan Project.* New York: Harper and Row.

Kennan, George F. 1967. *Memoirs, 1925–1950.* Boston: Little, Brown.

Lilienthal, David E. 1964. *The Atomic Energy Years, 1945–1950.* Vol. 2 of *The Journals of David E. Lilienthal.* New York: Harper and Row.

Millis, Walter, ed. 1951. *The Forrestal Diaries.* New York: Viking Press.

Stimson, Henry L. "The Henry Lewis Stimson Diaries." Stimson Literary Trust. Manuscripts and Archives. Yale University, New Haven. Photocopy at Truman Library, Missouri.

Truman, Harry S. 1956. *Years of Trial and Hope.* Vol. 2 of *Memoirs.* New York: Doubleday.

Williams, Francis. 1961. *A Prime Minister Remembers: The War and Post-War Memoirs of the Rt. Hon. Earl Attlee.* London: William Heinemann.

Newspapers

New York Times.
Times (London).
Washington Daily News.
Washington Post.
Washington Star.

Interviews

Arneson, Gordon R. Interview by Neil M. Johnson, June 21, 1989. Interview 456, transcript. Oral History Collection, Truman Library, Independence, Missouri.
Gordon, Lincoln. Interview by Richard D. Mckenzie, July 17, 1975. Interview 232, transcript. Oral History Collection, Truman Library, Independence, Missouri.
Osborn, Frederick. Interview by Richard D. Mckenzie, July 10, 1974. Interview 236, transcript. Oral History Collection, Truman Library, Independence, Missouri.

SECONDARY SOURCES

Books

Botti, Timothy J. 1987. *The Long Wait: The Forging of the Anglo-American Nuclear Alliance, 1945–1958.* New York: Greenwood Press.
Bullock, Alan. 1983. *Ernest Bevin: Foreign Secretary, 1945–51.* London: W. W. Norton.
Bundy, McGeorge. 1988. *Danger and Survival: Choices about the Bomb in the First Fifty Years.* New York: Random House.
Burridge, Trevor. 1985. *Clement Attlee: A Political Biography.* London: Jonathan Cape.
Bush, Vannevar. 1970. *Pieces of the Action.* New York: William Morrow.
Caute, David. 1978. *The Great Fear: The Anti-Communist Purge under Truman and Eisenhower.* New York: Simon and Schuster.
Cave Brown, Anthony, and Charles B. MacDonald, eds. 1977. *The Secret History of the Atomic Bomb.* New York: Delta/Dell.
Coit, Margaret. 1957. *Mr Baruch.* Boston: Houghton Mifflin.
Conant, James B. 1952. *Anglo-American Relations in the Atomic Age.* Oxford, England: Oxford University Press.
Costello, John. *Mask of Treachery.* 1988. London: Collins Sons.
Etzold, Thomas H., and John Lewis Gaddis, eds. 1978. *Containment: Documents on American Policy and Strategy, 1945–1950.* New York: Columbia University Press.

Feis, Herbert. 1957. *Churchill, Roosevelt, Stalin: The War They Waged and the Peace They Sought.* Princeton, N.J.: Princeton University Press.

Freeman, J. P. G. 1986. *Britain's Nuclear Arms Control Policy in the Context of Anglo-American Relations, 1957–68.* London: Macmillan.

Gowing, Margaret. 1964. *Britain and Atomic Energy, 1939–1945.* New York: St. Martin's Press.

———. 1974. *Policy Making.* Vol. 1 of *Independence and Deterrence: Britain and Atomic Energy, 1945–1952.* London: Macmillan.

———. 1974. *Policy Execution.* Vol. 2 of *Independence and Deterrence: Britain and Atomic Energy, 1945–1952.* London: Macmillan.

Harris, Kenneth. 1982. *Attlee.* London: W. W. Norton.

Hartcup, Guy, and T. E. Allibone. 1984. *Cockcroft and the Atom.* Bristol, England: Adam Hilger.

Helmreich, Jonathan E. 1986. *Gathering Rare Ores: The Diplomacy of Uranium Acquisition, 1943–54.* Princeton, N.J.: Princeton University Press.

Herken, Gregg. 1980. *The Winning Weapon: The Atomic Bomb in the Cold War, 1945–1950.* New York: Alfred A. Knopf.

Hershberg, James G. 1993. *James B. Conant: Harvard to Hiroshima and the Making of the Nuclear Age.* New York: Alfred A. Knopf.

Hewlett, Richard G., and Oscar E. Anderson. 1962. *The New World, 1939–1946.* Vol. 1 of *A History of the United States Atomic Energy Commission.* Pittsburgh: Pennsylvania State University Press.

Hewlett, Richard G., and Francis Duncan. 1969. *Atomic Shield, 1947–1952.* Vol. 2 of *A History of the United States Atomic Energy Commission.* Pittsburgh: Pennsylvania State University Press.

Holloway, David. 1994. *Stalin and the Bomb.* New Haven, Conn.: Yale University Press.

Isaacson, Walter, and Evan Thomas. 1986. *The Wise Men: Six Friends and the World They Made.* New York: Simon and Schuster.

Kaplan, Fred. 1983. *The Wizards of Armageddon.* New York: Simon and Schuster.

Lafeber, Walter. 1989. *The American Age: United States Foreign Policy at Home and Abroad since 1750.* New York: W. W. Norton.

Lamphere, Robert J., and Tom Shachtman. 1986. *The FBI-KBG War.* New York: Random House

McLellan, David. 1976. *Dean Acheson: The State Department Years.* New York: Dodd, Mead.

Melissen, Jan. 1993. *The Struggle for Nuclear Partnership: Britain, the United States and the Making of an Ambiguous Alliance, 1952–1959.* Groningen: Styx Publications.

Messer, Robert L. 1982. *The End of an Alliance: James F. Byrnes, Roosevelt, Truman, and the Origins of the Cold War.* Chapel Hill: University of North Carolina Press.

Morgan, Kenneth O. 1984. *Labour in Power, 1945–1951.* Oxford, England: Clarendon Press.

Moss, Norman. 1987. *Klaus Fuchs.* New York: St. Martin's Press.

Pincher, Chapman. 1987. *Traitors.* New York: St. Martin's Press.

Pringle, Peter, and James Spigelman. 1981. *The Nuclear Barons.* New York: Holt, Rinehart and Winston.

Rhodes, Richard. 1986. *The Making of the Atomic Bomb.* New York: Simon and Schuster.

Ruston, Roger. 1990. *A Say in the End of the World: Morals and British Nuclear Weapons Policy, 1914–1987.* Oxford, England: Oxford University Press.

Sherwin, Martin J. 1975. *A World Destroyed: The Atomic Bomb and the Grand Alliance.* New York: Alfred A. Knopf.

Walker, Martin. 1993. *The Cold War: A History.* New York: Henry Holt.

Wheeler-Bennet, John. 1962. *John W. Anderson, Viscount Waverly.* London: Macmillan.

Index

Acheson, Dean, 3, 5, 119; on Churchill-Truman summit, 193; on Fuchs affair, 169, 171–72, 173–74; integration proposal and, 146, 148–52, 154, 163; international control and, 104, 108; on Morrison, 180; on prior consultation, 177–78, 184; on Quebec Agreement, 144; on raw material allocation, 91, 92; relationship with Franks, 138; as replacement for Marshall as secretary of state, 142, 144; resignation, 121; on revelation of wartime agreements, 101–2, 116, 117; on security, 169, 192; on Soviet bomb possession, 157–58; on U.K. military intervention in Middle East, 181; on unawareness of atomic issues, 73, 80, 84–85, 89; on U.S. control of atomic weapons, 96; on U.S. military support for plants in U.K., 113–14; on U.S. raw material supply, 160

Acheson-Lilienthal Committee, 104–5

Adamson, Keith F., 16

Addison, Lord, 110

Advisory Committee on Atomic Energy, 75, 78, 79, 83, 110

Advisory Committee on Uranium, 9, 16, 19, 20, 22. *See also* National Defense Research Council (NDRC)

Agreement and Declaration of Trust (1944), 57

air plans, 193, 200. *See also* military

Akers, Wallace A., 27; attempt to discover U.S. reasons for noncollaboration, 42; on Conant, 52, 212n. 27;

connection to ICI, 44, 59; as director of scientific mission, 28, 33, 35; on Groves's security measures, 60; on independent British program, 45, 46; lobbying effort for collaboration, 37–38, 39; summons of British scientists to U.S., 57–58; visit to U.S. in 1942–43, 36–41

Alexander, A. V., 110, 137

Allibone, T. E., 125

Anderson, Clinton, 96

Anderson, Sir John, 23, 24, 75; acquisition of raw materials, 56–57; on Bohr, 67; on compartmentalization, 60; on Conant Memorandum, 52; concession to erect pilot plant in U.S., 28; Eldorado and, 56; on French Situation, 63, 64, 65; hesitance to commit to joint effort with U.S., 28; on McMahon Act, 100–101; on postwar interests, 49–50, 62; on replacement of Akers, 59; requests for collaboration, 33–34, 38; on restricted exchange, 40, 82; revision of Quebec Agreement and, 79–80, 81, 88, 89, 90, 93; on U.K. uranium and heavy water supplies, 46; on U.S. request for information, 42, 43. *See also* Advisory Committee on Atomic Energy

Anglo-American alliance, 1–3, 29–30; decline after Pontecorvo spy scandal, 174; formalization of, 24; full cooperation, 198, 202, 205; nonexistence, 187

Anglo-Egyptian treaty, 181

249

Baruch, Bernard, 104–7

Belgium: communists in, 108, 120–21; leak about integration and, 151; as part of "Western Union," 128; sale of uranium to Germany, 15; U.K. agreement with, 64, 79; uranium supply, 12, 56–57, 122. *See also* Congo

Berlin crisis, 127, 140–41

Bevin, Ernest, 75; on aid to Greece and Turkey, 112; on Baruch Plan, 106; on control of atomic weapons, 97; on cooperative discussions after Fuchs affair, 171–74, 177; on GEN 163 committee, 110–11; meeting with Byrnes about cooperation, 113; opposition to West German rearmament, 178; replacement of, 180; on Soviet threat to security, 140–41; on U.K.-U.S.-Canadian cooperation, 124, 126, 127–29; on U.S. military in U.K., 179

Bidault, Georges, 128

Blackburn, Raymond, 79, 106

Blackpool, conference at, 74

Blair House, 122, 123, 149–52, 233n. 23

Bohr, Niels, 10, 17, 66–67, 157

boiler project, 27, 32, 33

bomb, atomic: availability to U.K. in time of need, 158, 163–64, 200; collaboration on, 38, 39–40, 117; custody of, 142; disarmament, 105–7; doubts about production, 13; exchange for plutonium, 174; exchange for U.K. discontinuation of weapons manufacture, 137, 164; exchange for U.K. plutonium, 173, 186; exchange of information about, 97, 134–36, 139, 142, 200; German development of, 24, 26, 36; integration proposal and, 163; military importance of, 40, 44, 50, 67–68, 98–100, 134, 142, 149, 173, 204, 205; ownership of patent, 204; peaceful use of, 105; postwar collaboration and, 66, 88; prior consultation, 81, 117, 122, 126, 151, 175–78, 184, 191, 193; Soviet develop-

ment of, 6–7, 50, 97, 106, 115, 157–59, 168, 172; storage, 161–62, 163, 173, 200; U.K. production of, 46–47, 93, 118, 131–35, 137, 141, 144, 147, 153, 155–56, 180, 182, 184, 187, 188, 195–97, 199, 204; U.S. army responsible for development, 34; U.S. production, 126, 173

bomb, hydrogen, 160, 174; amended Atomic Energy Act and, 198; exchange of information on, 199; opposition to, 170; Soviet development of, 207; U.K. development of, 201, 202, 207; U.S. development of, 172, 188, 197, 200

bomb testing, 194–97

Bradley, Omar, 173, 179

Bragg, Lawrence, 12

Brazil, U.K. materials agreements with, 79

Briggs, Lyman, 16, 20, 22, 25

British Atomic Energy Research Establishment, 110, 139, 160, 167, 174

British Scientific Office, 19, 23

British Security Service, 167

British Supply Council, 19

Bundy, Harvey H., 49, 50, 66

Burgess, Guy, 5, 166, 169, 175, 181–82, 186, 206

Burns, James, 155

Burridge, Trevor, 219n. 30

Bush, Vannevar, 2, 4; on Anglo-Russian agreement, 42; Attlee-Truman summit and, 80–81; on Baruch, 105; on Bohr, 67; on British scientists in U.S., 33, 57–58; on collaboration, 23–26, 28, 33–35, 37–41, 44–47, 54, 58; on Conant Memorandum, 52; director of NDRC, 19, 20, 21–22; on Eldorado takeover, 56; on industrial interests, 59; influenced by Conant, 212n. 27; on information exchange, 122; on interim presidential policy committee, 218n. 7; international control and, 104; meeting with Perrin, 48–49; on Modus Vivendi,